DATA -ISM

Inside the Big Data Revolution

STEVE LOHR

ONEWORL

A Oneworld book

First published in Great Britain and the Commonwealth by
Oneworld Publications, 2015

First published in the US by Harper Business, an imprint of
HarperCollins Publishers

ISBN 978-1-78074-518-3
eISBN 978-1-78074-519-0

Printed and bound in Great Britain
by TJ International Ltd, Padstow, Cornwall

Oneworld Publications
10 Bloomsbury Street
London WC1B 3SR

To Terry and Nikki, for time served,
and all the laughs along the way

ACKNOWLEDGMENTS

A book may be written by a single person, but books are always collaborations. An editor nudges things along, a publisher makes a bet, and the adventure begins. There is an original idea, but it gets refined along the way, modified, honed and crafted by reporting, thinking and guidance.

The germ of this book really came in January 2012, with a call from a senior editor at the *New York Times* and former editor of the *Times* Magazine, Gerry Marzorati. At the time, he was helping guide the recently revamped Sunday Review section, and Gerry observed that I had written a number of articles about the field that was being called big data, and that the subject was getting more attention all the time, including being prominently on the agenda of the World Economic Forum annual meeting in Davos that month. We talked for a while, and he asked if there was a long overview piece to be done that explained the phenomenon for a general audience. I replied that I thought there was. The result was a lead article for Sunday Review that ran in early February 2012 under the headline "The Age of Big Data."

The Sunday Review piece generated a lot of interest, Web traffic, and reader comments. I hadn't thought of a book, but my agent, Max Brockman of Brockman Inc., did, and encouraged me, and offered sound advice. Hollis Heimbouch, publisher of the HarperBusiness imprint of HarperCollins, was the editor who made the bet that what is a rich subject could be made into a story—and that I could do it. Hollis is everything you want in an editor—by turns enthusiastic, patient, and insistent, someone who listens to your ideas and brings plenty of incisive ones of her own. She's also a very sure hands-on text editor in every way, from shaping the structure of a book to cutting. At HarperCollins, I'd also like to thank Eric Meyers, an associate editor and a deft practitioner of the dark arts of manuscript wrangling.

At the *Times*, several editors have—and still do—encourage my reporting in the field of data science, and its implications. They include Larry Ingrassia, Dean Murphy, Damon Darlin, Suzanne Spector, Joseph Plambeck, James Kerstetter, Quentin Hardy, David Gillen, Lon Teter, Thomas Kuntz, and David Corcoran. Conversations with Natasha Singer have sharpened my thinking on the subject of data and privacy. Then there is John Markoff, a science reporter these days; he and I have labored agreeably and often together in the data-laden precincts of business and science coverage for years.

While I was on leave from the *Times*, Mark Hansen generously offered me office space to work in and travel from and an appointment as a research fellow at the Brown Institute for Media Innovation. Mark is a director of the Brown Institute, a collaboration between the Columbia Graduate School of Journalism and the Stanford School of Engineering. Mark and Bernd Girod, the Stanford director of the institute, are leading an innovative effort to use data tools in storytelling and public policy. At Columbia, my workspace was in the career services office, and my thanks to the professionals there for kindly making the accommodation: Julie Hartenstein, Gina Boubion, Anusha Shrivastava, Izabela Rutkowski, and, earlier, Jacqueline DeLaFuente.

This is a book of narrative nonfiction. So I am especially thankful to the people who agreed to be interviewed, and showed the patience for follow-up and fact-checking inquiries. At the top of the list is Jeffrey Hammerbacher, who gave the most in time, thoughts, and candor. He not only went through many hours of interviews but also offered introductions to his parents, his wife, classmates from high school and college, and friends and work colleagues, past and present. All with no strings attached.

For other parts of the book, there were people I spent whole days with, such as Timothy Buchman at the Emory medical center in Atlanta and Nick Dokoozlian in the grape vineyards of central California and Michael Haydock in suburban Minneapolis.

Many others were interviewed for this book. They include Sam Adams, Brooke Barrett, Richard Berner, Patrick Bosworth, Thomas Botts, Erik Brynjolfsson, John Calkins, Murray Campbell, Dennis Charney, Herbert Chase, Jeffrey Chester, Sharath Cholleti, Adam D'Angelo, Arne Duncan, Sue Duncan, Tony Fadell, Edward Felten, David Ferrucci, Rachana Shah Fischer, Brian Gehlich, Jim Goodnight, Nagui Halim, Hendrik Hamann, Glenn Hammerbacher, Lenore Hammerbacher, Danny Hillis, Jeffrey Immelt, Jon Iwata, James Kalina, Kaan Katircioglu, Gary King, Jon Kleinberg, Martin Kohn, Randy Komisar, Patricia Kovatch, Edward Lazowska, and Michael Linderman.

They also include Mark Malhotra, James Manyika, Yoky Matsuoka, Andrew McAfee, David McQueeney, Douglas Merrill, Tom Mitchell, Craig Mundie, Arvind Narayanan, Tim O'Donnell, Sharoda Paul, Alex Pentland, Claudia Perlich, Stephanie Pieroni, William Pulleyblank, Tara Richardson, Virginia Rometty, Itamar Rosenn, Jeff Rothschild, Marc Rotenberg, Alex Rubinsteyn, William Ruh, Eric Schadt, Benjamin Scheuer, Neal Sidhwaney, Larry Smarr, Andrew Smeall, Jim Spohrer, Halle Tecco, Menka Uttamchandani, Veronica Vargas, Anil Varma, David Vivero, David Vladeck, Donald Walker, Danny Weitzner, and Michelle Zhou.

At IBM, I owe particular thanks to Jon Iwata and Ed Barbini for clearing the way for a broad reporting swath with IBM scientists and executives, and to Vineeta Durani and Angela Lee Sullivan for tirelessly making the interviews happen.

Finally, thanks to Terry and Nikki for understanding about work and the missed vacations, and to Nickie Trucco for my missing the Christmas holidays in 2013. But then, as I told them at the time, this book was a good problem to have.

CONTENTS

DATA
-ISM

HOW BIG IS BIG DATA?

J ust outside Memphis, an industrial symphony of machines and humans shuttles goods to and fro, their carefully orchestrated movements and identifying marks tracked by bar-code scanners and chips emitting radio waves. Mechanical arms snatch the plastic shrink-wrapped bundles off a conveyor belt, as forklifts ferry the packages onto trucks for long-distance travel. Flesh-and-blood humans guide and monitor the flow of goods and drive the forklifts and trucks.

McKesson, which distributes about a third of all of the pharmaceutical products in America, runs this sprawling showcase of efficiency. Its buildings span the equivalent of more than eight football fields, forming the hub of McKesson's national distribution network—a feat of logistics that sends goods to 26,000 customer locations, from neighborhood pharmacies to Walmart. The main cargo is drugs, roughly 240 million pills a day. The pharmaceutical distribution business is one of high volumes and razor-thin profit margins. So, understandably, efficiency has been all but a religion for McKesson for decades.

Yet in the last few years, McKesson has taken a striking step further by cutting the inventory flowing through its network at any given time by $1 billion. The payoff came from insights gleaned from harvesting all the product, location, and transport data, from scanners and sensors, and then mining that data with clever software to identify potential time-saving and cost-cutting opportunities. The technology-enhanced view of the business was a breakthrough that Donald Walker, a senior McKesson executive, calls "making the invisible visible."

In Atlanta, I stand outside one of the glassed-in rooms in the fifth-floor intensive care unit at the Emory University Hospital. Inside, a dense thicket of electronic devices, a veritable forest of medical computing, crowds the room: a respirator, a kidney machine, infusion machines pumping antibiotics and painkilling opiates, and gadgets monitoring heart rate, breathing, blood pressure, oxygen saturation, and other vital signs. Nearly every machine has its own computer monitor, each emitting an electronic cacophony of beeps and alerts. I count a dozen screens, larger flat panels and smaller ones, smartphone-sized.

A typical twenty-bed intensive care unit generates an estimated 160,000 data points a *second*. Amid all that data, informed and distracted by it, doctors and nurses make decisions at a rapid clip, about 100 decisions a day per patient, according to research at Emory. Or more than 9.3 million decisions about care during a year in an ICU. So there is ample room for error. The overwhelmed people need help. And Emory is one of a handful of medical research centers that is working to transform critical care with data, both in adult and neonatal intensive care wards. The data streams from the medical devices monitoring patients are parsed by software that has been trained to spot early warning signals that a patient's condition is worsening.

Digesting vast amounts of data and spotting seemingly subtle patterns is where computers and software algorithms excel, more so than humans. Dr. Timothy Buchman heads up such an effort at Emory. A surgeon, scientist, and experienced pilot, Buchman uses a flight analogy to explain his goal. GPS (Global Positioning System) location data on planes is translated to screen images that show air-traffic controllers when a flight is going astray—"off trajectory," as he puts it—well before a plane crashes. Buchman wants the same sort of early warning system for patients whose pattern of vital signs is off trajectory, before they crash, in medical terms. "That's where big data is taking us," he says.

T he age of big data is coming of age, moving well beyond Internet incubators in Silicon Valley, such as Google and Facebook. It began in the digital-only world of bits, and is rapidly marching into the physical world of atoms, into the mainstream. The McKesson distribution center and the Emory intensive care unit show the way—big data saving money and saving lives. Indeed, the long view of the technology is that it will become a layer of data-driven artificial intelligence that resides on top of both the digital and the physical realms. Today, we're seeing the early steps toward that vision. Big-data technology is ushering in a revolution in measurement that promises to be the basis for the next wave of efficiency and innovation across the economy. But more than technology is at work here. Big data is also the vehicle for a point of view, or philosophy, about how decisions will be—and perhaps should be—made in the future. David Brooks, my colleague at the *New York Times*, has referred to this rising mind-set as "data-ism"—a term I've adopted as well because it suggests the breadth of the phenomenon. The tools of innovation matter, as we've often seen in the past, not only for economic growth but because they can reshape how we see the world and make decisions about it

A bundle of technologies fly under the banner of big data. The first is all the old and new sources of data—Web pages, browsing habits, sensor signals, social media, GPS location data from smartphones, genomic information, and surveillance videos. The data surge just keeps rising, about doubling in volume every two years. But I would argue that the most exaggerated—and often least important—aspect of big data is the "big." The global data count becomes a kind of nerd's parlor game of estimates and projections, an excursion into the linguistic backwater of zettabytes, yottabytes, and brontobytes. The numbers and their equivalents are impressive. Ninety percent of all of the data in history, by one estimate, was created in the last two years. In 2014, International Data Corporation estimated the data universe at 4.4 zettabytes, which is 4.4 trillion gigabytes. That volume of information, the research firm said, straining for perspective, would fill enough slender iPad Air tablets to create a stack more than 157,000 miles high, or two thirds of the way to the moon.

But not all data is created equal, or is equally valuable. The mind-numbing data totals are inflated by the rise in the production of digital images and video. Just think of all of the smartphone friend and family pictures and video clips taken and sent. It is said that a picture is worth a thousand words. Yet in the arithmetic of digital measurement, that is a considerable understatement, because images are bit gluttons. Text, by contrast, is a bit-sipping medium. There are eight bits in a byte. A letter of text consumes one byte, while a standard, high-resolution picture is measured in megabytes, millions of bytes. And video, in its appetite for bits, dwarfs still pictures. And forty-eight hours of video are uploaded onto YouTube every minute, as I write this, with the pace likely to only increase.

The *big* in big data matters, but a lot less than many people think. There's a lot of water in the ocean, too, but you can't drink it. The more pressing issue is being able to use and make sense of data. The success stories in this book involve lots of data, but typically not in volumes that would impress engineers at Google. And while advances in computer processing, storage, and memory are helping with the data challenge, the biggest step ahead is in software. The crucial code comes largely from the steadily evolving toolkit of artificial intelligence, like machine-learning software.

Data and smart technology are opening the door to new horizons of measurement, both from afar and close-up. Big-data technology is the digital-age equivalent of the telescope or the microscope. Both of those made it possible to see and measure things as never before—with the telescope, it was the heavens and new galaxies; with the microscope, it was the mysteries of life down to the cellular level.

Just as modern telescopes transformed astronomy and modern microscopes did the same for biology, big data holds a similar promise, but more broadly, in every field and every discipline. Far-reaching advances in technology are engines of economic change. The Internet transformed the economics of communication. Then other technologies, like the Web, were built on top of the Internet, which has become a platform for innovation and new businesses. Similarly big data, though still a young technology, is transforming the economics of discovery—becoming a platform, if you will, for human decision making.

Decisions of all kinds will increasingly be made based on data and analysis rather than on experience and intuition—more science and less gut feel.

T hroughout history, technological change has challenged traditional practices, ways of educating people, and even ways of understanding the world. In 1959, at the dawn of the modern computer age, the English chemist and novelist C. P. Snow delivered a lecture at Cambridge University, "The Two Cultures." In it, Snow dissected the differences and observed the widening gap between two camps, the sciences and the humanities. The schism between scientific and "literary intellectuals," he warned, threatened to stymie economic and social progress, if those in the humanities remained ignorant of the advances in science and their implications. The lecture was widely read in America, and among those influenced were two professors at Dartmouth College, John Kemeny and Thomas Kurtz. Kemeny, a mathematician and a former research assistant to Albert Einstein, would go on to become the president of Dartmouth. Kurtz was a young maths professor in the early 1960s when he approached Kemeny with the idea of giving nearly all students at Dartmouth a taste of programming on a computer.

Kemeny and Kurtz saw the rise of computing as a major technological force that would sweep across the economy and society. But only a quarter of Dartmouth students majored in science or engineering, the group most likely to be interested in computing. Yet "most of the decision makers of business and government" typically came from the less technically inclined 75 percent of the student population, Kurtz explained. So Kurtz and Kemeny devised a simple programming language BASIC (Beginner's All-purpose Symbolic Instruction Code), intended to be accessible to non-engineers. In 1964, they began teaching Dartmouth students to write programs in BASIC. And variants of Dartmouth's BASIC would eventually be used by millions of people to write software. Bill Gates wrote a stripped-down BASIC to run on early personal computers, and Microsoft BASIC was the company's founding product. Years later, Gates fondly recalled the feat of writing a shrunken version of BASIC to work on the primitive personal computers of the mid-1970s. "Of all the programming I've done," Gates told me, "it's the thing I'm most proud of."

Back in the 1960s, Kemeny and Kurtz had no intention of making Dartmouth a training ground for professional programmers. They wanted to give their students a feel for interacting with these digital machines and for computational thinking, which involves analyzing and logically organizing data in ways so that computers can help solve problems. The Dartmouth professors weren't really teaching programming. They were trying to change minds, to encourage their students to see things differently. Today, when people talk about the need to retool education and training for the data age, it is often a fairly narrow discussion of specific skills. But the larger picture has less to do with a wizard's mastery of data than with a fundamental curiosity about data. The bigger goal is to foster a mind-set, so that thinking about data becomes an intellectual first principle, the starting point of inquiry. It's a mentality that can be summed up in a question: What story does the data tell you?

The promise of big data is that the story is far richer and more detailed than ever before, making it suddenly possible to see more and learn faster—or in the McKesson executive's words, "to make the invisible visible." And the improvement is not a little bit better, but fundamentally different. I think of this as the deeper meaning of Moore's Law. In a technical sense, the law, formulated by Intel's cofounder Gordon Moore in 1965, is the observation that transistor density on computer chips doubles about every two years and that computing power improves at that exponential pace. But in a practical sense, it also means that seemingly *quantitative* changes become *qualitative*, opening the door to new possibilities and doing new things. In computing, you start by calculating the flight trajectory of artillery shells, the task assigned the ENIAC (Electronic Numerical Integrator and Computer) in 1946. And by 2011, you have IBM's Watson beating the best humans in the question-and-answer game *Jeopardy!*

To a computer, it's all just the 1's and 0's of digital code. Yet the massive quantitative improvement in performance over time drastically changes what can be done. Trained physicists in the data world often compare the quantitative-to-qualitative transformation to a "phase change," or change of state, as when a gas becomes a liquid or a liquid becomes a solid. It is an apt, descriptive comparison. But I prefer the Moore's Law reference, and here's why. When the temperature drops below thirty-two degrees Fahrenheit or zero degrees Celsius, water freezes. It happens naturally, a law of nature. Moore's Law is an observation about what had happened for years, and what could well happen in the future. But it is not a law of nature. Moore's Law has held for so many years because of human ingenuity, endeavor, and investment. Scientists, companies, and investors made it happen.

The same is true of big data. It has become technically possible thanks to a bounty of improvements in computing, sensing, and communications. But the steady advance in software and hardware, and the rise of data-ism, will happen because of brains, energy, and money. The big-data revolution requires both trailblazing individuals and institutional commitment. The narrative of this book is built around one of each—a young man, and an old company. The young man is Jeffrey Hammerbacher, thirty-two, who personifies the mind-set of data-ism and whose career traces the widening horizons of data technology and methods. Hammerbacher grew up in Indiana, went to Harvard University, and then briefly was a quant at a Wall Street investment bank, before building the first team of so-called data scientists at Facebook. He left to be cofounder and chief scientist of Cloudera, a start-up that makes software for data scientists. Then, beginning in the summer of 2012, he embarked on a very different professional path. He joined the Icahn School of Medicine at Mount Sinai in New York, where he is leading a data group that is exploring genetic and other medical information in search of breakthroughs in disease modeling and treatment. Medical research, he figures, is the best use of his skills today.

At the other pole of the modern data world is IBM, a century-old technology giant known for its research prowess and its mainstream corporate clientele. Its customers provide a window into the progress data techniques are making, as well as the challenges, across a spectrum of industries. IBM itself has lined up its research, its strategy, and its investment behind the big-data business. "We are betting the company on this," Virginia Rometty, the chief executive, told me in an interview.

But for IBM, big data is a threat as well as an opportunity. The new, low-cost hardware and software that power many big-data applications—cloud computing and open-source code—will supplant some of IBM's traditional products. The company must expand in the new data markets faster than its old-line businesses wither. No company can match IBM's history in the data field; the founding technology of the company that became IBM, punched cards, developed by Herman Hollerith, triumphed in counting and tabulating the 1890 census, when the American population grew to sixty-three million—the big data of its day. Today, IBM researchers are at the forefront of big-data technology. The projects at McKesson and Emory, which will be examined in greater detail later, are collaborations with IBM scientists. And IBM's Watson, that engine of data-driven artificial intelligence, is no longer merely a game-playing science experiment but a full-fledged business unit within IBM, supported by an investment of $1 billion, as it applies its smarts to medicine, customer service, financial services, and elsewhere. The Watson technology is now a cloud service, delivered over the Internet from distant data centers, and IBM is encouraging software engineers to write applications that run on Watson, as if an operating system for the future.

The new and the old, the individual and the institution are at times conflicting forces but also complementary. It is hard to imagine that Hammerbacher and IBM would ever be a comfortable fit, but they are heading in the same direction—and both are big-data enthusiasts.

Another conflicting yet complementary subject runs through this book, and it centers on decision making. Big data can be a powerful tool indeed, but it has its limits. So much depends on context—what is being measured and how it is measured. Data can always be gathered, and patterns can be observed—but is the pattern significant, and are you measuring what you really want to know? Or are you measuring what is most easily measured rather than what is most meaningful? There is a natural tension between the measurement imperative and measurement myopia. Two quotes frame the issue succinctly. The first: "You can't manage what you can't measure." For this one, there appear to be twin claims of attribution, either W. Edwards Deming, the statistician and quality control expert, or Peter Drucker, the management consultant. Who said it first doesn't matter so much. It's a mantra in business and it has the ring of commonsense truth.

The second quote is not as well known, but there is a lot of truth in it as well: "Not everything that can be counted counts, and not everything that counts can be counted." Albert Einstein usually gets credit for this one, but the stronger claim of origin belongs to the sociologist William Bruce Cameron—though again, who said it first matters far less than what it says. Big data represents the next frontier in management by measurement. The technologies of data science are here, they are improving, and they will be used. And that's a good thing, in general. Still, the enthusiasm for big-data decision making would surely benefit from a healthy dose of the humility found in that second quote.

F or more than a decade at the *New York Times,* I have covered the technology ingredients and issues that now carry the "big data" label—well before the term entered the vernacular and became yet another unavoidable buzzword. And I still do. But this book is an effort to go both deeper and wider by surveying the projects and ideas on this frontier across the broader economy—and by talking to the individual scientists, entrepreneurs, and business executives who are confronting the technological and human challenges that data-ism inevitably creates. My reporting has been guided by the belief that if modern data technology is going to be a big deal economically, it has to go mainstream; it has to be deployed in almost every industry. The early triumphs of the consumer Internet—personalized search, targeted online ads, tailored movie recommendations, and the like—are impressive. But applying these technologies and techniques to huge industries of the physical world, like medicine, energy, and agriculture, is a more difficult challenge—and ultimately a more significant achievement, affecting far more people in far more ways. In the pages that follow, we will take a look at the progress of big data across the broader economy. We will be looking for the substance behind the salesmanship. Where is data-ism taking us? Where does big data shine, and where does it stumble?

2

POTENTIAL. POTENTIAL. POTENTIAL.

Jeffrey Hammerbacher is trying to win converts. He stands beside a lectern, pacing back and forth, addressing about a hundred people in an auditorium at the Mount Sinai medical school on the Upper East Side of Manhattan. Many in the audience wear the white lab coats of physicians. Hammerbacher has deep-set piercing eyes, an angular nose, a close-cropped beard, and a head of thick dark-brown hair. Brushing it into place is not always a priority. His title at the medical school is assistant professor of genetics and genomic sciences, but white lab coats are not his style. His shirts of choice are tight-fitting pullover jerseys or T-shirts. Both show off his brawny shoulders, thick biceps and forearms—the physique of the star baseball pitcher he was in high school; he still does his pitcher's weight workout a couple of times a week.

To this gathering of physicians and medical researchers, Hammerbacher delivers a brisk overview of his data tactics and philosophy. He runs through some of the basics of data handling: "instrument everything" you can with data-generating sensors; store all the data you can immediately, and figure out what to ask it later; make your data open to others in your organization, and let them experiment with it.

The practice of data-driven discovery, Hammerbacher observes, is just getting under way in most fields. Observation rather than prediction should be the near-term goal. "Before you can predict the future, you have to see the present," he says, adding that "seeing the present clearly can be as hard as predicting the future." Yet greater clarity, in the present or in the future, starts with data. "If you don't have the data, you can't do the science," he declares. "Data is the intermediate representation of science."

Months later, I am having lunch with Hammerbacher at a sandwich shop in New York. He is so intent on conversing between bites that he doesn't notice the spatter of mayonnaise on his cheek. "It's snobbery on my part," Hammerbacher says, "but I view math as the true arena in which human intellect is demonstrated at the highest level." His remark, both honest and telling, sticks with me. It's a mentality that speaks volumes, and not just about the young man having lunch with me. Data, the numerical facts of measurement and analysis, and the people most at home in the quantitative world are on the ascent. Increasingly, it seems, value and wisdom reside in data, the central tenet of data-ism. There is an arrogance to data-ism, but Hammerbacher, engaging, articulate, and uncommonly self-reflective, is its benevolent face. He even has an elegant term to describe its purpose, which is to tap "the numerical imagination," seeing beyond the bits of data and the numbers themselves to the underlying story they are trying to tell.

In his young career, Hammerbacher has scouted the frontiers of the data economy. On Wall Street, he was a "quant," building maths models for complex financial products. At Facebook, he started a team that began to organize and mine social-network data, for insights on how to improve the service and target ads. He called himself and his coworkers "data scientists," a term that was a curiosity then but has since become the hottest of job categories.

At Mount Sinai, he brings the same tool kit and mind-set. The goal, Hammerbacher says, is "to turn medicine into the land of the quants."

It may seem tempting to dismiss Hammerbacher's vision as the naïve hubris of youth. But there are intellectual elders who share his perspective, about medicine and beyond. Dennis Charney, dean of the Mount Sinai medical school, predicts that the explosion in genetic and health data combined with advances in computer analysis will bring a "paradigm shift" in medicine. It is the same story in science, sports, politics, and public health, and in industries as varied as advertising and agriculture. Gary King, director of Harvard University's Institute for Quantitative Social Science, calls it a "revolution" that is just getting under way, but one that "will sweep through academia, business and government. There is no area that is going to be untouched." The advance of data-ism is creating tensions and meeting resistance. "There is a war in every field about the interlopers from quantland," King says. Often, he explains, the schism results from people staring across a cultural divide of misunderstanding, of the false choice between fully human and fully automated decision making. King, as an emissary from quantland, says he offers an olive branch and cooperation. "My pitch," he says, "is, We're going to help you." Still, the drift of things seems clear. Alex Pentland, a computational social scientist at the Massachusetts Institute of Technology Media Lab, sees the promise of "a transition on a par with the invention of writing or the Internet."

The ranks of data scientists—people who wield their maths and computing smarts to make sense of data—are modest compared to the workforce as a whole, but they loom large. Data science is hailed as the field of the future. Universities are rushing to establish data science centers, institutes, and courses, and companies are scrambling to hire data scientists. There is a trend-chasing side to the current data frenzy that invites ridicule. But it is hard to argue the direction.

Jeffrey Hammerbacher was always a numbers kind of guy. His mother, Lenore, has kept a school paper he wrote as a seven-year-old. The assignment was to describe your favorite hobby. "My favorite hobby is doing math while I'm eating," Jeff printed in clear block letters. "I like doing this because math is my favorite subject and I like to eat." Then, one after another, he methodically enumerated the body parts and their function in his favorite activity of calculating while eating. "I need my tounge"—his one misspelled word—"to taste things and my brain to think and my lungs to breathe and my hands to hold the paper and pencil and also write." And on he went: his eyes to see the number problem, his teeth to chew the food, his stomach to digest it, and his heart to pump his blood. "by Jeff"

So the essence was evident early on—his affinity for the quantitative world and his instinct for scientific observation. Today, Hammerbacher is a thirty-two-year-old millionaire many times over. His wife, Halle Tecco, a Harvard MBA, is the founder and chief executive of Rock Health, which provides seed funding and advice for technology start-ups in health care. When they bought an apartment in Manhattan's East Village in 2013 from the actress Chloë Sevigny, both the *New York Post* and Britain's *Daily Mail* took notice. The *Post* called the pair "technology entrepreneurs" and the *Daily Mail* described them as a "tech power couple."

The milestones on Hammerbacher's life path suggest a sure thing. His is a résumé that speaks of ambition and likely privilege, of someone destined for success in the modern economy of money and technology. That would be a safe assumption for many, but it doesn't fit for him. Like so many of the most interesting people in any field, Hammerbacher is not the predictable product of his background, the data of his life.

Hammerbacher was born in Kalamazoo, Michigan, and his family moved to Fort Wayne, Indiana, when Jeff was five years old. His father, Glenn, a factory worker for General Motors, was transferred to the Indiana plant after cutbacks in Michigan, as Detroit was losing ground and market share to the more efficient automakers in Japan. The Fort Wayne plant would be the last of three GM factories for Glenn, who spent thirty-two years with the company, before retiring in 2004. Jeff's mother was a nurse who still does some volunteer nursing. After Glenn retired from GM, the couple eventually moved to South Carolina, taking the opportunity to flee the bitterly cold midwestern winters. Jeff's brother, Jay Ryan Hammerbacher, who is two years older, is a personal trainer—a growing occupational category in America's postindustrial service economy, if one far removed from his younger brother's high-tech niche.

Amid the turmoil in America's manufacturing sector over the last three decades, Glenn Hammerbacher was one of the lucky ones. His seniority in the United Automobile Workers union helped shelter him from the industrial storm in the Midwest. But Glenn also demonstrated the flexibility to pick up the family and move to new locations, and the skills to take on new jobs over the years. In return, he enjoyed a premium wage for factory work, ample health care coverage and benefits, and a pension untouched by the GM bankruptcy filing in 2009. Glenn's salary, supplemented by Lenore's income as a nurse, meant that the Hammerbacher family had a solid middle-class lifestyle. They lived in a two-story house on the outskirts of Fort Wayne with a swimming pool and a wooden deck, which Glenn built. The factory hours were predictable, and Lenore often worked part-time, so there was time to spend with the boys' activities like Cub Scouts and sports, and time and energy for hands-on parenting.

Even as a preschooler, Jeff had a mind of his own. When he was three years old, Lenore got a call one day from his teacher. Jeff was causing trouble, the teacher explained. One day, for whatever reason, he decided not to participate in any of the class activities. Worse, the other children were following his example, as if an incipient nursery school strike might break out at any moment. Lenore suggested that the teachers ignore him, and it seemed to work since the problem passed. But Lenore recalled the warning from the preschool teacher that Jeff would have trouble all his life if he didn't learn to cooperate and join in. It would be a recurrent theme over the years. Lenore had a little talk she used on him. People, she said, are good at different things. Maths was easy for him, but others struggled with it. He had to work on cooperating and conforming to get along in life. "He never bought it," Lenore recalls. Glenn sums it up: "If he didn't want to do something, he didn't do it."

Glenn and Lenore are informal, direct, no-nonsense Midwesterners—people of firm handshakes and straight talk. There was room for youthful rebellion in the Hammerbacher household, but only up to a point. Some things were not negotiable. Glenn and Lenore are devoted Roman Catholics, and that meant church every Sunday, in suit and tie, for the sons. "Never a question," Glenn notes. The parental lecture about life ambitions and values, as Glenn recalls saying, was: "I don't care if you're a garbage collector as long as you're the best garbage collector you can be." It may not have always been obvious to his parents, but the message got through. As he observed during one of our conversations, "My parents never valued being clever above working hard."

Jeff was plenty clever, though. When he started kindergarten he was still four years old, and after a few days the teacher called home. She said she thought that Jeff could already read. Yes, Lenore replied, but they didn't mention it openly at home because his older brother was not yet reading. "We didn't want to make a big deal of it," Lenore says. Even as a small child, he was an omnivorous reader, including all the directions that came with the boys' toys, which he stored in a box. His parents jokingly called it "Jeff's directions collection." Soon, he was one of the local library's best customers. Unless his parents stopped him, Jeff would often stay up most of the night, reading in his bedroom. One tactic that worked for a while was to stuff towels underneath the door so Lenore couldn't see the light inside.

When he was five or six, Jeff's answers to flash cards for addition, subtraction, multiplication, and division came so quickly that Glenn was suspicious. He couldn't help wondering whether Jeff had memorized the answers based on the order of the cards. But constantly shuffling the deck didn't slow him down. When he was seven years old, Jeff had his first Communion and received small amounts in cash and checks in envelopes from relatives and family friends. Glenn recalled Jeff thumbing quickly through the stack of envelopes and adding up the three hundred-some dollars in his head. "Some of the math things he did as a child kind of amazed me," Glenn says.

In high school, a maths instructor called home about a problem with Jeff's practice test for the Advanced Placement exam in calculus. The answers were correct, but Jeff did not show his work. If he did not write down the calculating steps that led to the solution, the instructor said, Jeff would be marked down. But to Jeff, writing down several intermediate steps was not necessarily the way he solved the problems. He is one of those people whose brains are wired for maths, who see equations and things sort of click in their heads; they do skip steps on the way to the answer.

Jeff was introduced to computer programming on an Apple IIe, when he was eight, in a summer school course. His first project was a space program, a pixelated UFO that could roam among some simple on-screen planets. The early programs were written in BASIC, but by middle school he had moved on to more advanced computer languages like Pascal and C++. Jeff was also an avid player of first-person shooter games like Doom and Quake, skillfully moving up the levels, zapping monsters and gathering more weapons as rewards. He also mastered software tools for getting onto the online gaming network DWANGO (Dial-up Wide-Area Network Game Operation) and the consumer online service AOL for free. "That's what you did if you were good with computers—you hacked into things," he recalls.

Jeff was a recalcitrant student, but also a gifted one. Starting in elementary school, Jeff was given special instruction in maths, pulled out of ordinary classes. In addition to public and Catholic parochial schools, Fort Wayne has a small private school, Canterbury School, and its high school is similar to the college prep schools on the East and West Coasts. Every year, Canterbury held an exam for those who could not afford the tuition. The prizes for the highest scores were scholarships to Canterbury. Jeff aced the test. This would be the story of his school years. Classrooms generally held little appeal. But he excelled at standardized tests, posting near-perfect scores.

In recent years, Canterbury has made an effort to become more diverse than Jeff recalls it in his days, as a school for Fort Wayne's affluent classes—the children of doctors, lawyers, business owners, and managers. "To me, they were fancy rich kids," Jeff says. He worked hard in his first year at Canterbury as a tenth-grader, earned A's and won academic awards, his mother recalled. But after his sophomore year, he lost interest. "He was done cooperating," Lenore says. In his meetings with Jeff's parents, the guidance counselor at Canterbury began with a mantra: "Potential, potential, potential." The point, of course, was that Jeff had plenty but he wasn't applying himself.

At Canterbury, Jeff was bright, a discipline problem, and a star pitcher on the baseball team. His friends, as he put it, were a mixture of "geeks and jocks." One friend was Rachana Shah. She was a grade ahead of Jeff, but they shared a couple of classes. In an advanced calculus class, they were the only two students. In their Latin class, there was one other student. In temperament, she was his alter ego, a type A personality, as she describes herself. But intellectually, she was a kindred spirit. The daughter of a local business owner and a physician, she lived in the moneyed Sycamore Hills section of town, beside a golf course. Indeed, Canterbury School in her years was "country club-ish," as she describes it. In that setting, Jeff was an exotic specimen, an autoworker's son with an attitude.

Today, Rachana Fischer (her married name) is a litigator in the Silicon Valley office of the law firm Paul Hastings. In high school, Jeff, she recalls, did "dumb show-offy things guys do," including drag racing on icy streets in winter and knocking down the light pole in front of the school. But he had read not only every maths book in the school, but also the work of poets like Frank O'Hara of the New York School and the Russian futurist Vladimir Mayakovsky. "He's a bit of a poet," Fischer says. And she was impressed by his raw intellect. "He made it look effortless," she recalls. "He could read a book in a day. It was not hard work for him."

The students at Canterbury School were governed by an honor code of honesty and adherence to the rules. It was a point of pride, for example, that there were no locks on the lockers at Canterbury. Fischer was the head of the student honor-code tribunal. One afternoon, the lab work in the chemistry class that both she and Jeff attended ran long. Next was their two-person calculus class. The heck with it, Jeff said, he was going to skip the class. It was a moral dilemma for the honor-code queen. If she skipped the class, it was a rule infraction. If she went to maths class, she would be asked about Jeff. "It was the first time I ever skipped a class," she recalls.

Fischer, a graduate of Harvard, and later Harvard Law School, wrote a recommendation for Jeff on his application to Harvard. The theme of her recommendation essay, she said, was that "achievement isn't always something that can be easily measured." She wrote that she thought of Jeff as a "true genius" even though that was "not reflected in his grade point average." In recent years, Fischer has followed Jeff and his professional success, and she is amused by the irony. "His career is based on analyzing people by data and numbers, but he doesn't fit into any box himself," she observes.

B aseball, as well as brains, got Jeff into Harvard. As a pitcher, he wasn't big, at six feet tall and 185 pounds, or overpowering. His fastball was good, but his curve was better. He did well enough in high school and summer leagues to get invited to group tryouts for major league baseball teams and to be recruited by colleges. The University of Michigan recruited him. But when Jeff and his father visited the campus at Ann Arbor, an assistant coach there, who had accepted a coaching post at Harvard, urged the Hammerbachers to consider Harvard, after seeing Jeff's high SAT test scores.

Jeff had not thought of Harvard before then, but the more he did, the more appealing it seemed—"a ticket out of the Midwest," as he put it. His baseball and his test-taking skill got him into Harvard, with enough financial aid so that tuition cost the Hammerbachers no more than if Jeff had gone to a state university in Indiana. When they dropped Jeff off on campus, Glenn and Lenore were not prepared for the coed dorms. "We almost pulled him out," Lenore says. Not really, Glenn adds, "But it was very liberal, and we're not liberal." And that only added to their misgivings about the university environment, where class attendance was not mandatory or monitored, meaning that Jeff would go to class only if he wanted to. And that wasn't often.

In his first two years at Harvard, Hammerbacher estimates that he showed up in a classroom no more than a dozen times. This was his period of extreme scholarly indifference, but classrooms were never his natural habitat. Teachers are authority figures, and he seems to be instinctively distrustful of authority, as if questioning assumed expertise as opposed to empirical evidence. This skepticism of authority has extended to bosses. His first real job was in high school as a cashier at Scott's Food and Pharmacy in Fort Wayne. One Friday evening, he wanted to leave early to watch the high school football game. The store manager threatened to fire him if he did. Hammerbacher walked out and got fired. Rather than tell his parents, he dressed up each day in his cashier's attire—black pants, black shoes, white shirt, and smock—and drove off to Scott's and parked his car in the lot. He then walked across the street to the public library, logging eight-hour reading stints. "I could spend my whole life doing that," he recalls with a smile. Lenore found out and had his car towed.

He resisted his parents' educational efforts as well. When Jeff was fifteen, his autoworker father thought it was time to teach him to change and rotate the tires on a car—a useful life skill. That lesson, Glenn recalls, ended after one tire, when Jeff stopped and declared, "Why should I do this? I have no intention of doing manual labor."

Hammerbacher displays a similar skepticism, and a certain arrogance, for intellectual authority in all of its forms, whether in classrooms or books. For example, when exploring a new subject, he applies a three-book rule: read at least three books from different perspectives, to "subtract out the author bias," as he puts it. His insistence on educating himself on his own terms—gathering his own data and exploring based on his interests—caused a lot of conflict during his school years. It irritated his teachers and perplexed his parents. "We thought he was just obstinate," Glenn says.

But Hammerbacher explains his approach as simply optimizing the learning machine that he is. He describes his preferred learning style as something close to self-designed tutorials. "I don't learn by listening to someone talk to me," he says. "I learn by reading and talking to people, interacting with them."

To say that Hammerbacher is a big reader is more than an understatement. His apartment in San Francisco has a two-story-high bookcase with more than 1,000 books, and he has 200 or so books on his Amazon Kindle. There are some overlaps—print and digital—but he has kept a spreadsheet of the books he has purchased and read since 2001. There are more than 1,100 entries. When I asked for a representative sample of books he's read and enjoyed, Hammerbacher sent a list of 299 books. It is a window into his mind, spanning intellectual history, economics, computer science, biology, philosophy, music, sports, fiction, and poetry. The list begins with *The Man Who Loved Only Numbers* by Paul Hoffman, and ends with *Infinite Jest* by David Foster Wallace. A few in between: *Relational Database Design* by Jan Harrington, *Anna Karenina* by Leo Tolstoy, *Modern Operating Systems* (2nd edition) by Andrew Tanenbaum, *Collected Poems* by James Dickey, *Artificial Intelligence: A Modern Approach* (2nd edition) by Stuart Russell and Peter Norvig, *Miles: The Autobiography* by Miles Davis with Quincy Troupe, *Principles of Economics* by

N. Gregory Mankiw, *To the Lighthouse* by Virginia Woolf. His list also includes four of J. K. Rowling's Harry Potter novels. A compulsive curiosity, his wife, Halle Tecco, explains, is "part of his DNA, and books are his avenue for learning."

At technical and industry conferences on big data, there is a lot of discussion of data scientists and what to look for in a really good one. The answer to the question typically boils down to two characteristics: a person who combines deep technical skill with a lively interest in the world beyond computing and maths. The need for technical skill is obvious, and the open-minded inquisitiveness is an asset because data science ranges across so many fields and relies on experimental discovery. Jim Spohrer, IBM's director of university programs, calls them T-shaped people, deep technically but also broad, with the top of the T being the breadth. T-shaped people, according to Spohrer, are more likely to be innovators and entrepreneurs not only in engineering but also in business, public policy, and academia. T-shaped people, Spohrer adds, are more skilled in the art of teamwork because they can communicate effectively with people in other disciplines. And Hammerbacher's reading list is a portrait of a T-shaped mind.

Yet Hammerbacher's first couple of years at Harvard were not the energetic pursuit of a distinctive, if unconventional, learning style. Partying proved a bigger attraction than study. Hammerbacher lost interest in baseball in his freshman year, quitting the team, and in the second semester of his sophomore year he just skipped finals altogether, failing all of his courses. His best friend had dropped out of school, and he had just broken up with his girlfriend. "Looking back now, it was very wasteful," Hammerbacher recalls. "I was a dumb kid."

Having flunked out of Harvard, and on his own, Hammerbacher managed to land a job with a small computer networking company in Queens, New York. He had a tiny apartment in the outer borough, but when he wasn't working, Hammerbacher spent a lot of time in Manhattan with Andrew Smeall over that summer. A Yale student at the time, Smeall was a high school friend of one of Hammerbacher's roommates at Harvard. Smeall recalls that Hammerbacher built a spreadsheet to guide his food purchases according to one standard: the most calories for the fewest dollars. The idea, Smeall says, was a "typical Jeff" project—by the numbers. Pound cake, peanut butter, and Goldfish crackers were staples of his diet. It seems an object lesson in the perils of maximizing for a single measurement—cost per calorie, in this case. The life lesson learned, in retrospect, Hammerbacher says: "Don't scrimp and save on food. It's where your mood comes from."

The penny-pinching apparently didn't extend to nightlife. The pair got along well that summer, Smeall says, in part because "we both had trouble saying no—to one more drink, to one more place to go. It was fun to have someone who could keep up with you in that sort of way, who had the social stamina.

"A lot of it was impulse control, or lack of it," recalls Smeall, a Chinese scholar with an MBA degree as well. "He's very different now. We both are."

The computer-networking job didn't last long. A casual approach to showing up for jobs on time and a know-it-all attitude exhausted the patience of the managers of small firm in Queens. Hammerbacher was fired after three months. Then he called home. Glenn and Lenore told him they would pick him and his things up, but he would have to rent the van. After paying for the van, Jeff was down to his last $40. Back in Fort Wayne, Jeff's parents charged him $100 a week, which they put in escrow for when he went back to school. They told Jeff he could go to Purdue or Harvard—that was up to him. But Lenore, especially, was insistent that he would go back to college.

Along with the tough-love stance, there was concern. Lenore did volunteer nursing at Matthew 25, a local charity that provides free medical and dental care to uninsured, low-income, and homeless people. Many of them were highly intelligent, she says, but unable to manage in the world, succumbing to bad habits and bad choices. Jeff's behavior at Harvard—drinking and partying and ignoring responsibility—had a familiar look to Lenore. He seemed unmoored.

But the months back in Fort Wayne were a settling-out period for Jeff. His friends were off at college, so not around. Hard partying in Manhattan and Cambridge were replaced by quiet dinners around the kitchen table in Fort Wayne. His days were occupied with routine jobs, first at Barnes & Noble and later a short stint at the GM factory. Jeff was only there for about a month, but, as Glenn put it, he "got a feel for an assembly line." It all proved motivational and enlightening. He would return to college, and not nearby in Indiana but back to Harvard.

Hammerbacher calls the forced hiatus from Harvard "a turning point in my life." Living on his own in New York and then back in Fort Wayne gave him an appreciation of ordinary working life and of his parents. "My relationship with my parents totally changed," he recalls. "It opened my eyes to all they had done and the sacrifices they had made for me." (Hammerbacher would later buy the retirement home in South Carolina for his parents, telling them, "Think of it as a good return on the investment of all your hard work.")

Returning to Harvard, Hammerbacher got his job back at the library, a condition of his financial aid, and attended classes more regularly. One was a small maths seminar on probability. The hands-on project was to write a software program for speech recognition. It seemed a good test bed for maths, since calculating probabilities and matching patterns in sound frequencies is crucial in speech recognition. Not incidentally, the instructor was Paul Bamberg, a cofounder of Dragon Systems, a commercial pioneer in speech recognition software. The programming involved tasks like implementing a fast Fourier transform algorithm, which converts time or space to frequency, and vice versa. The seminar was for students with serious maths muscles, and there were only five students in the class. One of them was Mark Zuckerberg, who would found Facebook a year later. But the two did not get to know each other at Harvard. "Zuck was socially awkward then, so we didn't talk much," Hammerbacher recalls.

Socially awkward is not a phrase anyone uses to describe Hammerbacher. Rude at times, certainly. He misses meals and meetings. E-mail messages will go unanswered for days. But his friends, his colleagues, and his wife attribute his behavior partly to a mind immersed elsewhere. It may come off as thoughtlessness, but he's full of thoughts, just not about the day-to-day conventions of social commerce. He may be self-absorbed, but not socially awkward.

By the standards of Silicon Valley techies, Hammerbacher is a social butterfly. "He's very extroverted, which is very unusual for someone who is that technical and into data," observes Adam D'Angelo, former chief technology officer at Facebook, who is chief executive of Quora, an online question-and-answer start-up. His extrovert gene and many interests, D'Angelo says, make "Jeff very good at connecting with all kinds of people."

At Harvard, Hammerbacher demonstrated his skill at social persuasion as a "punchmaster"—or recruiter—for Spee, one of the all-male final clubs at the university. Traditionally, these social and eating clubs were exclusive redoubts for the sons of prominent, moneyed families. They still are, but less so. Spee, founded in 1852, has a reputation for being a progressive club in its way. John F. Kennedy and Robert F. Kennedy were members, and Spee was the first club to admit an African American, in 1965. And less than a decade ago, an autoworker's son played a leading role in the club.

The job of the punchmaster—new members are called "punches"—is "a mixture of sales and winnowing," explains David Vivero, a Spee alumni. "And Jeff was good at both." Applicants were plentiful, so there was a lot of meeting people face-to-face to quickly assess if a young man would be a good addition and, if so, try to persuade him to join the club. The goal, Hammerbacher explains, was a certain diversity, to assemble a collection of people with complementary interests and talents. "Spee," he says, "had a real cross-section of members socioeconomically, racially and otherwise, or as much as Harvard was."

Hammerbacher built a spreadsheet of the prospects, identifying the attributes of each, and with the other punchmaster put together a punchbook with candidates' pictures and data. He liked recruiting and the mutual courtship involved. "You're meeting new people, and there is a certain volatility and vulnerability on both sides, and an opportunity to form bonds," he explains. The maths whiz relished the unpredictability of the human encounter. Later, Hammerbacher would prove to be an adept recruiter of technical talent, at Facebook and Cloudera, and most recently the data team at Mount Sinai.

Throwback institutions like final clubs evoke a spectrum of reactions—from being aghast to being bemused to saying "why not." Hammerbacher's views reside on the latter, pragmatic side of the spectrum. When he returned to Harvard, most of

his friends were in final clubs (sophomores and juniors are recruited). At home and in sports teams, male environments were
the norm, so he was not uncomfortable in the single-sex social
setting. "I recognize all the issues, and it's pretty messed up on
one hand," he says. "But for me, the pros were stronger than the
cons."

The circle of friends at Spee was one brick in a more stable life
for Hammerbacher at Harvard. His time away had been a
learning experience. A more mature, motivated student returned,
determined to take advantage of the educational richness of the
place. His undergraduate thesis was in the field of machine learning, a branch of artificial intelligence that focuses on software
programs that can learn from data. He recalled one application
of machine learning in a course in computational neuroscience.
In an experiment, sensors were attached to a rat's head to capture
the pattern of synapses firing in its brain while it was swimming.
The assignment was to predict where the rat would swim next,
during the second half of its time in a tiny pool, based on studying
the pattern of brain signals during the first half of the rat's swim.
The course helped inspire a continuing interest in neuroscience—
beyond rodent brains to how the human brain works, and the
mystery of why even brilliant brains sometimes go awry. His was
an intellectual interest that was also deeply personal, as we'll see
later.

It took Hammerbacher a little longer than most, but he graduated from Harvard on a spring day in 2005, in early June, with
the weather cooperating and temperatures in the mid-sixties.
Glenn and Lenore came out from Indiana for the graduation cer-
emony, pleased and relieved. Shortly after they returned home,
Lenore wrote a brief note to Jeff's former guidance counselor at

Canterbury School. It began, "Potential, potential, potential." She couldn't resist.

H ammerbacher likes to describe his career as a matter of repeatedly following the smartest people in search of the best problem. His first stop after he graduated was finance. "All these math people who I felt were a lot smarter than I was were going to Wall Street," he recalls. In 2008, Bear Stearns collapsed and was sold in a fire sale to JPMorgan Chase, a prelude to the full-blown financial crisis that erupted later that year. Yet in the summer of 2005, Bear Stearns was thriving and hiring. The investment bank's headquarters occupied a new Madison Avenue office tower, and its pulsing moneymaking engines were its eight trading floors, each several times the size of a basketball court.

Hammerbacher was a quantitative analyst, working mainly with the bond and mortgage traders. He built stochastic models, which applied probability theory to calculate the likelihood that, say, interest rates, currency exchange rates, or mortgage-default rates would rise or fall.

The model makers do not supply the assumptions about the world, like economic growth and financial market trends. Economists, researchers, and traders do that. A model codifies the assumptions and is used to run millions of simulations to calculate the probability—or risk—of a movement, even a momentary one, in the price of a financial asset. They are used to build complex financial products. One that Hammerbacher worked on was a "double-barrier knockout derivative."

The model-making, Hammerbacher says, was "hard-core math" and "a lot of fun." The quants had office cubes on a floor above the traders, where they sat at desktop computers, designing and refining the algorithms that animated the models. The quants programmed in C++, but the results were converted from code to spreadsheets and printouts for the traders and sales people. The quants were informally dressed eggheads by Wall Street standards, who often wore sneakers, while the traders and sales staff sported the full-dress uniform—suits, ties, and dress shoes.

The trading floor, Hammerbacher recalls, had row after row of open work areas, each trader surrounded by six screens, working two phones, and keeping an eye on the MSNBC financial news programming on a television monitor, with suit jackets off and sleeves rolled up. The traders were paid the most and ultimately made the big-ticket decisions on the financial products and market bets—the top of the Wall Street food chain.

Given his family background, Hammerbacher was more sensitive than most to just how rarified a slice of the nation's economic life is Wall Street, and the concentration of wealth it generates. "I spent my days there sitting next to very large pools of capital," he says. "On a single mortgage hedge, we made or lost more in a day than my father made in a lifetime."

Hammerbacher stayed at Bear Stearns less than a year before moving on. He did not leave because of any particular misgivings about Wall Street. But he decided to take his quantitative talents elsewhere. Better problems were on the horizon. An episode on the trading floor one day got him thinking. Suddenly, the data feed—price quotes, bid and ask offers, financial news, everything—into the trading floor stopped. Whether because of a technical snafu or a human lapse, the data pipe was knocked out for a couple of hours. Activity on the trading floor ground to a halt.

Hammerbacher had been happily modeling away, day in and day out, but none of that work mattered without the data. The data, he decided, was more important than the models. And Wall Street, it seemed to him, had it wrong. "Our whole goal was to make the models more complex," he recalls. "It was a really bad use of quantitative skills."

The day the data stopped at Bear Stearns left a lasting impression. Hammerbacher calls it an "educational moment," a lightbulb switching on. It marked a shift in his thinking toward what he and others call a "data first" approach to knowledge discovery and decision making—start with the data, this digital power tool for observing and understanding.

His "educational moment" on Wall Street was merely an intimation of the importance of data, a starting point of a journey into dataland. He wanted to learn more. Reading would help. There was much to be learned from the rich history of data storage and analysis, and the evolution of artificial intelligence. But to really make progress, Hammerbacher decided he would have to go work for one of the Internet companies, born on the Web, which were becoming natural laboratories for data science.

Google was the leader in recognizing and exploiting the opportunity. It was a pure big-data company. Google applied the data-first strategy to both its search engine and its advertising business, and had created new software tools to do so. But there were other companies, including Internet stalwarts, like Yahoo! and Amazon, and fledgling start-ups. In early 2006, Hammerbacher joined one of the start-ups. It was just two years old and had fewer than fifty employees. Still, it had promise, as well as several people Hammerbacher knew from Harvard. It offered both adventure and a certain familiarity. So he moved to Silicon Valley and went to work for Facebook.

3

BET THE COMPANY

Data should be the vital raw material that strengthens and improves the machinery of decision making. Data is the input, and the output is smarter choices and wiser judgments. But the data paradox is that a world richer and richer in data has so far yielded little payoff in most fields. The supply of data races ahead, while the ability to use it lags badly.

To see the data paradox writ large, let's return to the intensive care ward at the Emory University Hospital in Atlanta. The term "information overload" dates back at least to the 1960s, and the futurist Alvin Toffler popularized the phrase in his 1970 best seller *Future Shock*. Yet it seems a quaint understatement to Dr. Timothy Buchman as he surveys the twenty-room intensive care unit, where the hundreds of electronic medical devices are throwing off 160,000 data points a second. In the glassed-in rooms, there is so much medical machinery, so many respirators, pumps, kidney machines, and monitors for every vital sign and major organ, that it can be hard to see the patient. "In the center of all that is a quivering mass of protoplasm—a human being," he says, pointing to the patient, a middle-aged woman. "That's important to never lose sight of, but it's getting harder and harder." There is a rationale, and a good one, for each device and each data stream, but the superabundance of information is stressful, distracting, and potentially hazardous in a high-pressure setting where doctors and nurses make one hundred decisions a day per patient. "We've got great people, but we're asking them to do incredible mental gymnastics," observes Buchman, director of the Emory Center for Critical Care.

T he data deluge that Buchman sees daily is partly a by-product of progress in medicine. People in intensive care units these days are generally sicker and older than they were a decade or two ago. Many people treated today with outpatient care would have been admitted to hospitals years ago, those in the general wards would have been in intensive care, and many of those treated successfully in intensive care today would have been dead. You can do more, so you do more, and technology plays a big part.

But the result is an overwhelming flood of data—numbers, trend lines, sounds—on those screens and devices. Each machine typically is its own electronic island, monitoring a specific organ or measuring one vital sign, and not "talking" to the other machines. The overwhelmed nurses, working twelve-hour shifts, are reduced to ricocheting from machine to machine, trying to keep up. "Nurses tend to focus more on the technology and less on the patient," says Stephanie Pieroni, a supervisor in the ward. "There's all this technology. It's just too much." In 2011, the data overload became so apparent that a "watcher" position was created. The watcher scans two computer screens in the center of the ward, monitoring a handful of basic vital signs, including heart rate, blood pressure, and oxygen saturation, on all twenty patients. It's another set of eyes to catch missed signals, guarding against "human error" in an environment that seems beyond human capability—especially for the front line of care. It's a stopgap step, a Band-Aid for the problem, but no more.

The humans need help. Just as technology creates the data paradox, technology has a major role to play in overcoming it. And Buchman is hoping to do just that in critical care medicine, a truly challenging testing ground. He's guiding a research project that could affect how medicine is practiced in intensive care units and beyond. His research is supported by the National Institutes of Health and involves collaboration with IBM's research labs and Excel Medical Electronics, a technology company that specializes in handling data from medical devices.

B uchman is both imposing and genial in person, a big man, with a shaved head and glasses, and a deep voice. He is a surgeon and a professor. He was raised in New York, was educated at the University of Chicago, and worked and taught at Johns Hopkins University and Washington University in St. Louis before coming to Emory in 2008. Buchman thoroughly enjoys teaching and finds it irresistible. I show up at his office, and he begins with a lengthy, and fascinating, history of critical care medicine, the concepts and the people. He begins in 1923, with Walter Dandy, a surgeon at Johns Hopkins, who had the idea of putting "cohort patients"—those at greatest risk—in a single room. Next stop in his historical sketch is Boston in 1942, and the Cocoanut Grove nightclub fire, which killed 492 people. He credits a young surgeon at Massachusetts General Hospital, Oliver Cope, with setting up the world's first burn unit. Cope and his colleagues set standards of care that included intravenous fluid resuscitation with saline solutions and dressing wounds with soft gauze and petroleum jelly. The contribution to critical care, Buchman says, was "the idea that a common practice of treatment could be delivered by teams of people."

Buchman, like most good teachers, often reaches for a metaphor to help explain things. To him, most medical data is the equivalent of an unexploited natural resource. "It's like being able to drill for oil and pull it out of the ground, but that's as far as we are," he says. The data oil, he adds, has not yet been refined to its version of kerosene and gasoline. Nor, he says, do we have the data counterparts to the internal combustion engine, automobiles, and roads—the elements of a complementary ecosystem of use and value.

Buchman's favorite metaphor is flight. In his office, Buchman has four flat-panel computer screens on his desk. Before he begins to show the work on his project, he brings up a website showing the air-traffic-control view of flights near Atlanta's Hartsfield-Jackson airport. For each plane in the air, there is a tiny icon of an airplane. Beside each plane icon, in a text box, is the flight's call signals, altitude, speed, and destination. Buchman's own plane, which he rents out when he's not using it, appears on the screen. The radar image shows that it is at 3,000 feet and climbing, traveling at 122 knots (about 140 miles per hour), and heading south. Multiple streams of data, from flight plans to Global Positioning System information, go into the simplified image on the screen. The air-traffic controller, or pilot, Buchman notes, doesn't need to wrestle with the underlying data. That, he says, is the better future for intensive care units—technology that shows the patient's position and where that patient is headed, a medical GPS system.

Buchman's critical care project doesn't have its polished GPS system yet. But what I see on Buchman's screens is progress toward that kind of visual presentation of data: a series of graphs for each patient, monitoring different health conditions with results rendered as patterns of green dots (safe territory) and red dots (trouble ahead). One graph shows the heart monitoring for atrial fibrillation, an irregular heart rate that can be a precursor to heart failure or stroke. Most of the patient graphs on Buchman's screen have mainly green dots, but one patient's graph is showing more red dots—an early warning that treatment is needed. Another data set shows neurological readings of intracranial pressure. A patient with a healthy heart rate and oxygen level in the blood is showing a red-dot spike in pressure in the brain, suggesting a high risk of hemorrhage.

Buchman is the field marshal of the project, but the dot patterns on the screen depend on technical experts, like Sharath Cholleti, who apply machine-learning algorithms to the streams of patient data. The data science challenge in medicine, Cholleti says, is more difficult than in finance or the consumer Internet. "Yes, it's going to be challenging in health care, but it can be done," he says. (Cholleti continues to work part-time on the Emory research, but in 2014 became the chief data scientist at CleMetric, a big-data start-up in health care.)

An intensive care unit, as its name suggests, is the epitome of high-stakes medical care, where decisions with life-and-death consequences must be made quickly and frequently. With all of its digital gadgetry, an intensive care unit is also the most data-rich environment in a hospital. That makes critical care, Buchman says, an ideal place to field-test the technology for parsing streams of real-time data and applying clever software for alerts and to ask what-if questions. If, for example, you give a particular drug to a certain patient, what is that patient likely to look like, biologically, in five or ten minutes. "In the past, we could only make retrospective decisions about what was right for a certain group of patients," Buchman says. "Now, we are getting very close to being able to make judgments about what is right for that particular patient, right now. That's the transformational force." Buchman is one of the leaders of a community of researchers trying to do predictive medicine in critical care settings, both in adult and neonatal wards for premature babies, including teams headed by Dr. Carolyn McGregor at the University of Ontario Institute of Technology and Dr. J. Randall Moorman at the University of Virginia.

N agui Halim, an IBM research fellow, explains that he is always on the lookout for "industry leaders who have transformational problems." And Buchman, he says, fit the bill. Based at the company's research lab in Yorktown Heights, New York, Halim leads a team that has developed a technology, called System S, for analyzing streams of data in high volume, like all of that data flowing from the Emory intensive care unit. The streaming software, Halim says, allows for continuous analysis, in minute detail, in a way a human never could. The technology never tires—a digital form of "constant vigilance," he notes. In critical care, the technology holds the promise of literally saving lives, addressing a "holy grail problem," as Halim put it.

Tackling daunting problems is part of the essence of IBM, home to a research staff of 3,000 at a dozen labs. There are, of course, the headline-generating public demonstrations of recent years—IBM's Deep Blue supercomputer defeating the world chess champion Garry Kasparov in 1997 and Watson's outwitting the best human *Jeopardy!* players, led by Ken Jennings, in 2011. And every day, IBM's technology is used for everything from simulating the global climate to plumbing the mysteries of life in molecular biology. Solving problems on the frontiers of science and technology is important to IBM, adding to its storehouse of intellectual assets and know-how. It also burnishes the image of IBM as an innovator, for a company that has no consumer products to show off its technology, unlike Apple, Google, and others. But what really sets IBM's corporate heart aflutter is problem solving that carries with it big, lucrative markets.

Big data is such a market, and IBM recognized that years before the term became popular. The notion that computers should let you do something smart with data has been around since the dawn of computing. More recently, software programs for using data to make better-informed decisions were given the label "business intelligence." It is a predecessor term to big data, and one still in use.

Business intelligence tends to focus on collection, reporting, and basic analysis but not on the predictive or experimental features of data science. The concept dates back to 1958, presented in a paper titled, "A Business Intelligence System," by Hans Peter Luhn, a computer scientist at IBM. Such a system, according to Luhn, was needed to cope with the postwar data boom in business, government, and the sciences. "Information," he wrote, "is now being generated and utilized at an ever-increasing rate because of the accelerated pace and scope of human activities." Luhn envisioned what would, decades later, become commonplace: coding documents with snippets of software, or tags, so they could be more easily sorted and "read" by a computer; and software for searching documents for the frequency of the appearance of words, suggesting patterns and signaling the relative importance of particular subjects and ideas. The purpose, he wrote, was to digest and present information to guide decision making. "Special equipment," he conceded, would be required. Luhn closed with a prescient bit of qualified optimism. "Perhaps the techniques which ultimately find greatest use will bear little resemblance to those now envisioned," he wrote, "but some form of automation will ultimately provide an effective answer to business intelligence problems."

BM's headquarters is so nestled in woods and a rocky ravine that you don't see it until well after you have entered the corporate grounds. The lazy Z–shaped building looks deceptively small from the outside. It is a striking contrast to the headquarters it replaced, a suburban corporate big house on a hill, opened in 1963. The new headquarters, far smaller, opened in 1997, makes a different statement, tucked into nature rather than rising above it. The exterior is mostly stainless steel, aluminum, and glass, while the interior has a lot of exposed wood and stone. The workspaces are a collection of open-plan, small conference areas and offices. It's all very modern, though some of the hallways have artifacts of IBM's heritage, early computers and calculators, even a Hollerith punched-card tabulator.

Down another hallway, in a small conference room, I sat down for a conversation with Virginia Rometty, IBM's chief executive, in early 2014. She is a tall woman, over six feet tall in heels. She's wearing a gray suit, and her blond hair is meticulously coiffed in a shoulder-length bob. On her right wrist is a slender black Fitbit band for tracking activity and sleep patterns. Her mornings start in the gym. At her alma mater, Northwestern University, Rometty majored in computer science and electrical engineering, and she was also president of a sorority. The combination says a lot about her. She's socially adept—good in business settings and with corporate customers, an essential skill to climb the management ranks at IBM. But she also has an affinity for the quantitative perspective. "I've always believed that most solutions can be found in the roots of math," she says.

Rometty has been a champion of the company's big-data strategy, long before the term became popular—first as a leader of the company's services business and later as the senior vice president in charge of strategy as well as sales. Since 2005, IBM has invested more than $25 billion in the data business, a large portion of that to buy dozens of companies that specialize in data analytics.

The pace of the data strategy has been ramped up since Rometty became chief executive in 2012. Forty thousand IBM consultants, engineers, sales people, and scientists working in the data business are spread across the company's services, software, and research divisions. In early 2014, Rometty announced that its prototype projects with the Watson technology in health care and other industries were sufficiently encouraging to justify creating a new business division. IBM will invest $1 billion in the Watson business and the unit would grow to 2,000 people. Watson has become a "cloud" software service, delivered Google-style over the Internet from remote data centers. IBM is sharing Watson technology with outside software developers and start-ups, so they can write applications that run on top of Watson. IBM has created a $100 million equity fund to jump-start that third-party development by outsiders. The company hopes that Watson can become the equivalent of an operating system for artificial intelligence software. The goal is to nurture a flourishing commercial ecosystem, in the way that Apple's iOS and Google's Android have done in smartphone software, and Microsoft's Windows did in personal computer software.

In early 2014, IBM also focused the activities of the 3,000 researchers in its labs to make data projects the priority. The realignment, according to Rometty, is the most significant shift at IBM research since the 1990s, when retooling for the Internet era became the imperative. A program of fundamental research, as in the materials science of computer hardware, will continue. Along with cloud computing, research will be concentrated on big-data projects in specific industries and the underlying machine-learning technologies used to find answers and insights in data, as Watson does. IBM refers to these machine-learning capabilities as "cognitive" computing. Pick your term, Rometty says—big data, analytics, or cognitive—but it's all in the same data neighborhood, and it is the direction in which IBM is unequivocally headed. "We are betting the company on this," she insists.

The bet makes sense today, Rometty says, because the technology and the economics have finally come together. Using data to make better decisions is both doable and affordable across a range of industries, realizing a long-sought promise of computing. "I feel I've worked on this my whole life," observes Rometty, who joined IBM in 1981. In the mid-1990s, she worked with the maths department of IBM research to create a unit called the Insurance Research Center. Its mission was to use data and algorithms to identify previously unrecognized attributes that seemed to make some people better insurance risks than others, even in the same demographic groups.

But insurance actuaries resisted, positive they knew their business far better than some maths nerds from IBM. In the data-first world, you have to be open to accepting what is in the data, which includes new and unknown relationships and correlations; things you never thought of. That lesson learned nearly two decades ago, Rometty says, rings true today. "Culture will be the biggest issue," she says. The insurance research group disbanded, and the IBM mathematicians went off to other errands.

Rometty decided to try again nearly a decade later, in 2004. By then, key technologies had matured further and data was proliferating in volume and variety. But IBM had changed as well. In 2002, it bought the consulting arm of the accounting firm PricewaterhouseCoopers for $3.5 billion. That purchase brought thousands of consultants with expertise in specific industries. It meant IBM was not just a technology supplier selling computer hardware and software, but a company that could combine technology and industry know-how to solve business problems for its corporate customers.

The lure for IBM was higher profit margins and a competitive edge over its technology company rivals. "We felt that the future was going to be to live at the intersection of business and technology," Rometty explains. Adding industry experts also changed the conversation when IBM approached companies. It was no longer just a bunch of computer salesmen and maths geeks trying to tell them how to improve their business, but IBM also had people who had toiled for years in their industries. And Rometty had a plan for blending IBM's technical and industry expertise to tackle real-world projects that would demonstrate the payoff from using data and clever software in business and government.

I n the fall of 2004, William Pulleyblank, a mathematician and computer scientist at IBM's Watson lab, was deep into super-computing. In fact, the machine he had worked on for years, Blue Gene, was about to win the title as the world's fastest su-percomputer, surpassing NEC's Earth Simulator of Japan. The IBM computer was designed for large-scale simulations of micro-biological phenomena, notably protein folding, and to break new ground in supercomputer design. Hence, its name—Blue, as a nod to the corporate nickname Big Blue, and Gene for its assigned task.

By 2004, IBM scientists were not surprised by occasional calls from corporate managers. It was a decade after IBM's near-death experience in the early 1990s, when the spread of inexpensive computing using personal computer technology hammered the company's traditional mainframe business. IBM famously recov-ered, led by Louis V. Gerstner Jr., the only outsider ever to serve as chief executive in the company's more than century of history. In the turnaround, Gerstner slashed bureaucracy and established a closer working relationship among the labs and the business units, the scientists and the suits. Given his background, Pulleyblank was more comfortable than most researchers working with in-dustry. A Canadian, Pulleyblank had shuttled between academia and business. His résumé included everything from developing computer models for logistics planning at the Canadian Pacific railway to teaching at the University of Waterloo for eight years, before joining IBM's research division in 1990.

Rometty, then a senior executive in IBM's services group, called him with a proposal: lead a start-up venture inside the company. The unit would tap people from the labs, the services group, and the software business. Its goal would be to build a business around helping companies and governments use the modern data flood to cut costs, boost sales, and streamline operations. The new venture would reside in the services group and be called the Center for Business Optimization. Pulleyblank hesitated at first, as a researcher with no experience running a business. But the more he thought about it, the more appealing the offer looked. "This was the idea I joined IBM to see if I could help make happen," Pulleyblank recalls. And he was encouraged and surprised that IBM's top executives picked him to head the new venture. "I thought to myself, sweet Jesus, the inmates are going to run the asylum," he recalls.

Optimization, in this context, is a big word to describe using data and computing to make the best decisions possible. That has been IBM's promise since the punched-card days, and there have been landmark projects along the way, like the company's collaboration in the 1950s and 1960s with American Airlines to pioneer computerized reservations, with its system called Sabre, an acronym for Semi-Automated Business Research Environment. (Spun off as a separate company years ago, Sabre is the reservations engine used today by hundreds of thousands of travel agents and hotels, and hundreds of airlines and online booking services, like Expedia and Travelocity.)

But more than a decade ago, IBM was seeing advances in technology and the rise of both more and new types of data that signaled a fundamental change in what could be done. That is, the company saw one of those thresholds at which quantitative change becomes qualitative, opening a path to a huge new potential market for IBM—and for competitors as well.

Every year, IBM's research labs produce a report, called the *Global Technology Outlook*, a headlights look into the near future for the company's management team. The 2003 outlook featured a section with the lengthy headline, "Data Explosion Drives New Information Organization, Retrieval, Analysis and Storage Paradigms." It discussed the rapid increase in real-time, or streaming, data from sensors and video cameras, and all of the Internet-era data like Web pages, click streams, and social-media messages. It pointed toward the coming impact of "ubiquitous connectivity" enhanced by more and more mobile devices, though Apple would not jump-start the smartphone market with the iPhone until 2007. IBM's senior executives are close readers of the annual technology outlooks, but the reports are not business documents. An aborning technology trend does not necessarily mean that a big market is around the corner, or a market that IBM, which caters to corporate customers, should pursue. But the "data explosion" report struck a chord with top management, they came up with a plan, and Rometty made her call to Pulleyblank.

Pulleyblank's center was soon up to about 100 people, expanding and working on projects over the next few years. In Stockholm, a road-use toll system deployed sensors, including cameras to photograph license plates, and analysis software. The rush-hour pricing—higher tolls in peak-traffic periods to discourage driving—and the monitoring technology cut roadway traffic by 25 percent, added 40,000 riders a day to mass transit, and reduced pollution. In Singapore, the IBM team analyzed data on car accidents and traffic to predict where traffic accidents were most likely to occur. The results were used to change traffic management practices to ease traffic bottlenecks and remove road hazards to prevent mishaps. In Norway, radio-frequency tags were attached to chickens and poultry products, tracking them from the barnyard to supermarket shelves to ensure freshness. In New York State, the IBM software combed through claims data to spot anomalies and curb Medicaid fraud. Other data-analysis projects boosted the efficiency of corporate tasks like fine-tuning manufacturing and transportation operations to reduce inventory costs at several companies. The work, Pulleyblank recalls, was driven by customer requests—companies, and government agencies asking for help—rather than being determined by IBM's research agenda or product plans. "And the real challenge," he says, "was to show you could make a business out of it, not just solve some one-off problem for a client."

As the proving-ground projects accumulated, they suggested an answer to an internal problem: IBM was suffering from something of an identity crisis. Inside and outside the company, no one seemed to have crisp answer to the simple question, "What is IBM?" The company had sold off its personal computer business, its lone consumer line. It was no longer mainly a mainframe computer company. It had large and growing businesses selling technology services and software to corporations, but nothing to capture the imagination there. In the spring of 2008, IBM sent summer interns with video cameras out on the streets of New York to ask passersby the question, "What is IBM?" They got shrugs and mumbles in reply. The exercise merely confirmed what IBM's formal market research studies showed. "We were mostly defined by what we were no longer," Jon Iwata recalls.

At the time, Iwata, senior vice president for communications, had just been handed the responsibility for marketing as well.

With the job came a mandate from IBM's chief executive, Samuel Palmisano, to do something about the company's identity deficit. For inspiration, Iwata dipped into the corporate archives. Mostly, he read the writings of Eliot Noyes, the architect and industrial designer who was hired by Thomas Watson Jr., son of the founder and IBM's president from 1952 to 1971. Noyes guided an overall corporate design program for IBM's buildings, interiors, logo, and products. He commissioned leading architects and designers including Eero Saarinen, Mies van der Rohe, and Charles Eames. Noyes himself designed the IBM Selectric typewriter, with its distinctive "golf ball" rotating type head. Iwata describes his research as "brand archeology." Yet those early documents, he adds, never mentioned the word "brand." Noyes, according to Iwata, was the "curator of IBM's corporate character."

In the 1960s and 1970s, IBM was esteemed and revered publicly, Iwata notes, without a presence in the consumer market (it entered the personal computer business in 1981). But it was known as a company, he says, that "invented things that changed the world," from the bar code, which enabled mass-market consumer brands and national retail chains, to the 360 mainframe computer, the information engine of control for postwar corporations. In Pulleyblank's projects, Iwata saw a similar theme of ushering in big changes, in transportation, food distribution, policing, and energy conservation—"stories of transformation," he put it. The common thread was that these were big societal and business problems that, in Iwata's phrase, "yield to data." He saw an opportunity to explain the meaning of the company as a leader in the evolution toward a data-driven world.

So there was a concept. But what words to put to the vision? For that task, Iwata tapped Michael Wing, a senior staff member with a knack for distilling the head-scratching complexity of technology into words that resonate with ordinary people. Wing and Iwata had long conversations about what memorable phrase could be the conceptual umbrella for what they were trying to communicate. They discussed "intelligent infrastructure," but that seemed too geeky. Iwata asked Wing to write a short essay, as if a speech by Palmisano, IBM's chief, to "sketch out thinking." The mock speech began with an overview of the impact of the Internet and globalization, "making the world simultaneously smaller and flatter." Next, Wing wrote, something else was under way that "may ultimately have an even greater impact on our lives and our future. The world is becoming not just smaller and flatter, but also *smarter*." The speech-like essay was titled, "A Smart Planet."

A couple of weeks later, in July 2008, during his regular monthly one-on-one meeting with Palmisano, Iwata showed him the essay. After reading the first few paragraphs, Palmisano said, "This is it." A week later, Iwata presented the idea to IBM's seventy-five top executives, who had gathered at the headquarters in Armonk, New York, for a two-day meeting. His presentation had thirty-three colorful PowerPoint slides, giving an overview of the technology trends and walking through some of the projects that Pulleyblank's team had worked on. Iwata finished with IBM's self-interest. It played to IBM's strength as a big and broad technology company that could act as a general contractor on such ambitious projects, supplying everything from research to chips. To outsiders and IBM employees, it would provide an answer to that question, "What is IBM?" It would be the company "building a smarter world."

The IBM executives seemed to like the idea, and so did an utterly unscientific sampling of family members. Apparently, civilians outside the technology industry got the message. That was enough for Palmisano, who didn't want to wait to launch the corporate identity and marketing campaign. "We did no market testing," Iwata concedes, a real departure from standard practice at IBM.

Once the go-ahead decision was made, IBM called in Ogilvy, its lead advertising agency, to prepare for a major marketing push in the fall. There was briefly some debate about the headline name, with the leading alternative being "Thinking Planet." But Smarter Planet it stayed. Ogilvy was hard at work on ad copy and the logo. But in September, after Lehman Brothers collapsed, the financial crisis hit with a vengeance. The global economy was reeling, and there were fears of a rerun of the Great Depression of the 1930s.

Palmisano was scheduled to give a talk at the Council on Foreign Relations in New York in early November, intending to use the occasion to present the Smarter Planet concept and kick off the marketing campaign. He planned to make a case for technological enthusiasm and a bright future just around the corner at a time when the global economy was facing its darkest hour in decades. His plan ran the risk of being greeted as a tone-deaf marketing pitch. He called Iwata into his office in October, wondering whether to shelve the campaign. As Palmisano put it, "I could look like a fool." Yet he and Iwata also considered another perspective. It could be a very good time for a dose of what Palmisano termed "fact-based optimism." They would sleep on it and decide in the morning. The next day Palmisano had decided, saying simply, "Yeah, let's go."

His talk at the Council on Foreign Relations never mentioned IBM, but it clearly had a self-serving side. Palmisano's message was received with skepticism and some furrowed brows. Yet the Smarter Planet campaign proved to be a success and tackling customers' problems that "yield to data" has become a big business for

IBM. If anything, the marketing campaign has been too success-ful, creating awareness for the power of data-parsing technology and lifting the sales of rivals.

One of the largest is SAS Institute, a commercial pioneer in data-analysis software. The company's origins date back to 1976, when it was founded by James Goodnight and three colleagues at North Carolina State University's agricultural statistics depart-ment. Its software was initially used to calculate the intricacies of soil, weather, seed varieties, and other factors to improve crop yields. Today, SAS sells its software to most Fortune 500 com-panies, employs more than 13,600 workers, and generates sales of nearly $3 billion. A lanky, laconic billionaire, Goodnight, the chief executive of SAS, says, "I think Smarter Planet is great. IBM gins up the demand, and we get the sales."

Over a breakfast at IBM's headquarters, Iwata admitted that there is a kernel of truth in that critique. The Smarter Planet cam-paign, he says, may have featured too many case studies about the problems of the planet, like the environment, and too few everyday corporate challenges. Still, he portrays that as an issue of tactical fine-tuning that has been addressed. "Mostly," Iwata says, "we got it right."

M ostly, IBM did get it right. Pan back, and what you see is the larger story—the evolutionary pathway of a bundle of signifi-cant, data-fueled technologies into the marketplace, from original research to pilot projects to a splashy public bet on an emerging business. Without advances in technology, there is no progress. But technology alone isn't destiny by any means. For technol-ogy to really pay off, other crucial ingredients must go into the recipe: investment, time, and the optimism to keep going, to keep spending the dollars and energy to make things happen. At its best, marketing is a believable narrative—a case for "fact-based optimism," as Palmisano put it.

The irony is that Palmisano's decision couldn't really be described as a data-driven decision. By 2008, there was an accumulation of demonstration projects that were encouraging, and some of the investments to build the business had been made, but the bulk of the investment would come later. Yes, IBM had done research on market and technology trends—all data points, in their way. But there was plenty of uncertainty, and this was the kind of decision that points to the limits of the big-data approach. Big data is good at *interpolation*—figuring out what happens next when the outcome is, most likely, a continuation of the current trend. It is far less good at *extrapolation*—figuring out what happens next when the trend line of the future is less clear. In short, big data stumbles when a decision requires an intuitive step outside the data sandbox—beyond the range of the data.

In 2011, as Palmisano was departing from IBM, with Rometty becoming chief executive in 2012, I asked him about the Smarter Planet decision more than three years earlier, and its timing just after the financial crisis struck. He said he had talked to IBM's scientists and gathered data as much as possible. Then, he made the bet, based on data but also intuition. "It was a judgment call," he explained. "If it didn't take judgment, a computer could do it."

In 2014, IBM altered the Smarter Planet marketing campaign. There would be less grand vision—that point had been made—and more examples of customers using modern data technology, as mainstream examples of data success have proliferated. Progress had been made since 2008. It was time to evolve again, and not just the marketing message. The century-long history of IBM is a story of adaptation, sometimes a smooth one, sometimes a struggle.

But the direction is clear, shifting to higher-profit businesses while shedding less profitable lines. "I think we've become pretty fearless about that," Rometty observes. A month before we talked, she had sold off IBM's business that makes less expensive data-serving computers, powered by industry-standard chips from the personal computer industry, for $2.3 billion to the Chinese computer maker Lenovo. The low end of the server computer business had become fiercely price competitive, and IBM brought no technical advantage to that market. It was not IBM's kind of business, even though it generated yearly sales of $4 billion.

The big-data era is the next evolutionary upheaval in the landscape of computing. The things people want to do with data, like real-time analysis of data streams or continuously running machine-learning software, pose a threat to the traditional computer industry. Conventional computing—the Von Neumann architecture, named for mathematician and computer scientist John von Neumann—operates according to discrete steps of program, store, and process. Major companies and markets were built around those tiers of computing—software, disk drives, and microprocessors, respectively. Modern data computing, according to John Kelly, IBM's senior vice president in charge of research, will "completely disrupt the industry as we know it, creating new platforms and players." IBM, of course, sees opportunity in that disruption. Its new Watson business, delivering machine-learning computing as a cloud service over the Internet, is one example. Another is the kind of real-time data stream analysis IBM's researchers are doing at the Emory medical center. But IBM also has large traditional software and hardware businesses that face the very threat that Kelly describes. And a sizable swath of IBM's services business involves engineers writing applications, using traditional software, for corporate customers. Today's big-data applications typically use cloud-style computing in which processing and software are delivered remotely, from distant data centers, over the Internet.

Under Rometty, IBM is making huge investments in the future—big-data technology and cloud computing. But the dilemma facing the company is whether the new business will grow faster than the old business erodes. In early 2014, when I spoke to Rometty, she talked of the lessons she had learned about the imperative of constant corporate evolution. "Don't fight cannibalization," she says at one point. Trying to preserve the past is a formula for failure, she notes, a lesson IBM learned the hard way in the 1990s. And the company's long history, she says, moving from one chapter to the next, time and again, affords perspective. "You kind of get used to that idea," Rometty observes, "and it's fairly liberating that you don't have to stay wed to the past."

A couple of months later in 2014, I spoke to Rometty again at corporate headquarters. She admitted that it was a "rocky time" for IBM, but necessary to prepare the company for the future. "We are transforming this company for the next decade," she says. "That is not a one-year job, not when you're a hundred-billion-dollar company."

Rometty has no doubts about the big bets she has made and that the direction is the right one. Her one concern, she says, is about the pace. "We are making progress," she says, "and we just need to keep moving with speed." But apparently it wasn't fast enough. In the fall of 2014, IBM abandoned its profit target for 2015, amid declining revenue and earnings. The new businesses showed strong growth, but they were still not big enough. The pace would have to pick up.

Yet getting ahead of the technological curve and trying to create a culture of continuous evolution is hard work. The task can be especially daunting in a huge corporation like IBM. Rometty needs a cadre of credible evangelists, and one of the most convincing is Kerrie Holley, an IBM technical fellow and a designer of cloud computing software. He holds several patents, and also a law degree. He remembers old software projects as if they were war stories. In 1986, he says, he was leading a fifty-person team assigned to design and install a new software system for Wells Fargo Bank, after it bought Crocker National Bank, linking the teller networks of both banks. He worked eighty-hour weeks and knew every computer language and software program involved—COBOL (Common Business Oriented Language), CICS (Customer Information Control System), IMS (Information Management System), and Assembler. "If my team couldn't do it, I could," he recalls.

Holley is not just a code horse, but a bit of a clothes horse as well. When we spoke at IBM's Almaden Research Center in San Jose, where the dress code tends toward jeans and T-shirts, Holley wore a tailored brown dress shirt, matching slacks, and oxford shoes, freshly polished. He is a fine-featured, light-skinned black man with a trim physique.

Holley's message to his IBM colleagues, he says, is that data-fueled software delivered over the Internet cloud poses a threat to "the very fabric of what we do" in the company's software and services businesses. The needed overhaul is partly technical, but also in the business model. Customers increasingly want to pay for how much they use—software by the sip—instead of writing multimillion-dollar checks for company-wide software licenses. The applications must also be tailored for mobile devices, smartphones, and tablets. The good news, Holley says, is that if IBM can retool its software for the new era, "we can really breathe life into the applications." That is his encouraging pitch for the "cloud first" approach that IBM wants to take across its software development labs worldwide.

Inside IBM, Holley is trying to foment cultural change in big successful businesses, encouraging them to adapt so they remain successful. His message is personal as well as strategic. Thriving businesses add workers, while faltering businesses cut jobs. Every year, thousands of workers are hired and thousands fired, across the company's global workforce of more than 400,000. The corporate euphemism for the annual rite is "workforce rebalancing." IBM does this more humanely than many companies, with comparatively generous severance payments and counseling. But there is a Darwinian feel to the annual exercise, an institutional recognition that no job is really safe. Holley's mission, he says, is to help people from "falling into the trash bin of their own experience."

Holley certainly avoided that pitfall in his own life. His earliest memory of his mother, Shereline Pitts, was when he was four years old, he recalls, and his grandmother took him to visit his mother in Joliet Prison, southwest of Chicago. "My mother led a life of crime," he says. She was a heroin addict who engaged in prostitution, check fraud, and other nonviolent crimes to support her habit. She was in and out of prison, Holley recalls, and the longest she ever lived with him was about six months. He grew up on Chicago's South Side, raised by his grandmother, along with his half brother and half sister, Laurencc and Lynette, both of whom are now dead. Holley recalls his grandmother, Martha Pitts, as a stern woman of limited education and expectations. "Her dream for me was to become a mailman," he says.

H olley's boyhood neighborhood was a tough one of gangs, guns, and drugs, and his grandmother, concerned for his safety, mostly kept him in the house after school. But when he was six years old, she let him go to an after-school program that offered free tutoring, the Sue Duncan Children's Center. He went there every day, and it became his home in a lot of ways, broadening his horizons and ambitions. About fifty people, Holley recalls, showed up at the center most days, ranging from kindergartners to high school students. College students from the nearby University of Chicago and older high school students would tutor the younger students. Holley began as a pupil and ended up as a teacher, even into his twenties during his college years and later, when he could. There was the education, but also basketball games and life lessons from Sue Duncan. "Finish what you start—she was a stickler for that," Holley says.

Sue Duncan recalls a young man who thrived on the structure and discipline of the program, and he was good at numbers. When he was thirteen, Holley was handling the accounting records for the center. His persistence, hard work, and character, she says, made him a role model for others. "Kerrie and I got along super," she recalls. "We helped each other out." With one gesture, she had a major influence on his education. When it was time for him to enter high school, she used her own address instead of his on his high school admission application. As a result, Holley would go to Kenwood Academy High School, on the edge of Hyde Park, a short distance but a world apart from the high school in his neighborhood. At first, he was put in remedial classes, but within two weeks he was placed in higher-level courses. Maths was his favorite class. He liked the numeric problem solving, he excelled, and maths, he says, was the "one subject where the teacher had no subjectivity in grading, which I also liked."

At the after-school center, Holley typically oversaw a small group of younger boys. One of them, who began at the center when he was five, was Sue Duncan's son, Arne. Holley tutored the group in reading, maths, and vocabulary. "He was my teacher every single day," recalls Arne Duncan, secretary of education in the Obama administration, who is ten years younger than Holley. His mother, Duncan says, treated Holley almost like a son. When he was eighteen, Holley got his two front teeth knocked out, and Sue Duncan simply wrote his grandmother a check for the dental work. Arne Duncan recalls Holley as a teacher, especially helpful when he was wrestling with algebra in the sixth grade. But he also recalls going out into the neighborhood around the center with Holley, sometimes to play basketball in the playgrounds, under the protective wing of the young man he came to know well. "Outside of my family, he was the most important formative influence on my life," Duncan says. His experience at his mother's inner-city center, witnessing the potential impact on underprivileged young people, Duncan says, steered him toward a career in education. "How many Kerrie Holleys are there in the country? Young people with amazing potential that no one gives a damn about," Duncan observes.

Perhaps many, but the arc of Holley's life is remarkable—proof, again, that extraordinary people so often cannot be explained by their backgrounds. No algorithm or data could predict their outcome.

4

SIGHT AND INSIGHT

S ensing, seeing, and acting. That is the promise of big data, and it seems straightforward. The improved measurement and monitoring that big-data technology makes possible are a kind of *seeing*. You can see things more clearly, and that clarity should improve the odds of intelligent action. But achieving that virtuous progression of data-driven decision making is anything but easy. Most companies can't see their operations very clearly. So much of the near-term opportunity for big data is not in fancy artificial intelligence software but in the more mundane realm of counting, monitoring, and seeing things with greater clarity.

The McKesson drug distribution case—$1 billion less inventory, an efficiency gain of roughly 13 percent—is a dollars-and-cents success story in big data. The idea of applying modern data science to a complex product distribution network originated with Kaan Katircioglu, an IBM research scientist at the time, who is an expert in operations research. "If you go to business school, they call it management science," he explains. "But it's all about numbers, really, bringing mathematics to the decision-making process and to business operations."

For years, Katircioglu says, his discipline suffered from a poverty of good data, but no longer. It is becoming possible to build a sophisticated digital model of a company's operations, and then run simulations to predict the likely outcomes of different management decisions. He had worked on the concept, but to get funding for a first-of-a-kind project Katircioglu had to find a company that looked like a good candidate and then convince the company to be the real-world guinea pig.

Katircioglu, helped by IBM's industry consultants, settled on McKesson. The first meeting was in 2009, and initially the IBM side seemed to be making a Smarter Planet pitch focusing on sustainability and reducing carbon emissions by optimizing transportation routes. But it soon became clear to the McKesson executives that the larger opportunity was being able to capture a "big picture" view of their business. In consultant circles, it is called the "end-to-end" view.

It is instructive to explain why McKesson was a promising test bed. McKesson is not just any company. It is, Katircioglu observes, "a well-oiled machine," efficient and focused. McKesson was early to embrace Six Sigma, a system of statistical measurement and methods for eliminating product defects and streamlining business operations. It invested heavily in scanning, sensor, and software technology. Starting in the 1990s, McKesson invested heavily in foundational information technology. First came the migration from paper to digital records and tracking, with the adoption of bar-code scanning and radio tags on products. The second was the implementation, completed in 2005, of software for computerizing operations, a so-called enterprise resource planning system, supplied by the giant in that market, SAP of Germany.

Those two engines of digitization generated much of the data that IBM's software then mined and modeled. But McKesson's data was not merely abundant but also "very clean," in Katircioglu's phrase. (In the spring of 2014, Katircioglu left IBM to join a company with truly abundant data, Google, as a quantitative analyst.)

Many companies live in a digital Tower of Babel, a hodge-podge of incompatible computer systems and data formats added over the years. Not McKesson. The company's distribution network produces massive amounts of data, but it comes from pretty tame sources—its sensors and its shipment-tracking software, for example. What is being measured—pills, prices, and shipment miles—translates easily into numbers. There is not a lot of randomness in the McKesson data. Its data was plentiful, stable, and reliable, which provided sturdy building blocks for IBM to fashion its facsimile of McKesson's operations—a flight simulator for decision making.

The result, according to Donald Walker, senior vice president for distribution operations, is that McKesson can now see its business far better at two levels. First, it can dig down into the data and determine its profit or loss by product, supplier, and customer. "We were never able to do that before," he says. "We never had that kind of granular analysis."

The second kind of seeing is seeing into the future. The technology is a tool for modeling what-if decisions, using the digital replica of the physical world to peer into the future to make more accurate predictions and better decisions. Take the example of very expensive drugs, typically cancer or specialized antibiotics drugs, whose cost can run into the thousands of dollars for a month's treatment. These drugs, in economic terms, have two main characteristics: they are quite valuable, and the demand for them is highly uncertain. So McKesson kept stocks at a few of its distribution centers and shuttled them around as needed. It amounted to a "best guess" approach.

Yet the McKesson executives wondered if there wasn't a better way. If these drugs were all kept in the suburban Memphis hub and air-shipped to customers, what would the effect be? What would be the impact on overall costs and delivery times? But there was natural resistance to the centralized approach. To a transportation manager, focused on curbing transport expenses, air shipments looked like an extravagance, costing about ten times more than shipping the drugs by truck.

But the IBM modeling software showed that the centralization plan should pay off, at least for one set of drugs its algorithms had identified. McKesson tested that prediction with a pilot project and found that inventory levels for the costly drugs were cut in half, more than making up for the higher airfreight expenses. On-time delivery for the drugs increased to 99 percent from less than 90 percent. The software model gave McKesson the clarity and the confidence to go ahead and is now being used to experiment and improve performance across the company's distribution network

Moreover, IBM is adapting the software it developed for the giant pharmaceutical distributor—a kind of SimCity for supply chains—to other industries. And while it may be exceptional in some respects, McKesson's success illustrates where the big-data approach shines today. It shows data really being used to guide decisions and to make better decisions, ones that trump best guesses and gut feel, experience and intuition.

The basic drift of data-ism seems unassailable: decisions of all kinds should be increasingly made based on data and analysis rather than experience and intuition. Science prevails; guesswork and rule-of-thumb reasoning are on the run. Who could possibly argue with that? But there is a caveat. Experience and intuition have their place. At its best, intuition is really the synthesis of vast amounts of data, but the kind of data that can't be easily distilled into numbers.

I recall a couple of days spent with Steve Jobs years ago, reporting a piece for the *New York Times Magazine*. Decisions that seemed intuitive were what Jobs called "taste." An enriched life, he explained, involved seeking out and absorbing the best of your culture—whether in the arts or software design—and that would shape your view of the world and your decisions.

One afternoon, we went to Jobs's home in Palo Alto. Several days earlier, he had hosted a small gathering there for Bill Clinton, who was then the president. The living room was still set up as it had been for the presidential visit, nearly empty except for a ring of wooden chairs, American craft classics. The chairs, Jobs observed, were George Nakashima originals, and he then offered a brief account of the Japanese American woodworker's life. Nakashima had a cross-cultural blend of experience, studying architecture, traveling on a free-spirited tour of the world, and working in different cultures. His designs were original, Jobs said, because Nakashima had a distinctive sense of taste, shaped by his life experience.

Steve Jobs was no quant, but he was an awesome processor of non-numerical data, curious, self-taught, and tireless. His real talent, as Jobs himself described it, was seeing "vectors" of technology and culture, where they were headed and aligning to create markets. And as a product-design team leader, he was peerless. Jobs worked at making it all appear effortless, even instinctive. In early 2010, for example, Apple's iPad tablet computer had been announced, but it was not yet on sale. Jobs came to the New York Times Building in Manhattan to show off the device to a dozen or so editors and reporters around the company's boardroom table. An editor asked how much market research had gone into the iPad. "None," Jobs replied. "It's not the consumers' job to know what they want." That, he suggested, was the job of his intuition.

Jobs may have been a high-tech product genius, but his intuition was not magic. It was the consequence of the experience and knowledge he had built into a sophisticated mental model of the world and how it works. That ability to mentally construct a model of the world that adapts with experience is a hallmark of human intelligence. A closer look at the strengths and weaknesses of human cognition provides clues to the role that data science can play in helping humans. In short, what is the sensible division of labor in decision making between man and machine?

I n the fall of 2013, IBM held a symposium at its Watson research lab that probed that issue, in the context of computer software that keeps getting smarter and smarter. The first speaker was Daniel Kahneman, the Princeton psychologist and Nobel Prize winner in economics. His 2011 best seller, *Thinking, Fast and Slow*, describes his research, with Amos Tversky, a mathematical psychologist, into the basis of common human errors. Humans, Kahneman explains, are often error-prone when we make decisions by applying rules of thumb and biases. The culprit is "fast" thinking, the quick assessment of a situation to take action. It is thinking anchored in experience, and seems easy—reading a novel, recognizing emotions in others, or the back-and-forth of social conversation. "Slow" thinking is the hard, synapse-burning kind—taking an intelligence test, calculating a tax filing, or writing a report.

Fans of Kahneman's book will recall that fast thinking—also "system-one" thinking, in his nomenclature—received a humbling critique. And in his talk at IBM, he revisits the indictment. We're overconfident, we oversimplify, and we suffer from my-side bias, which Kahneman defines as "anything that favors my opinion I am more sensitive to." We value the quality of the story more than the quality of the evidence. Fast thinking is very weak in statistical reasoning. To explain, he uses one of his research examples, plucked from his book. The research subjects are told that there is a man named "Steve," and he is described as "meek." Is he more likely to be a librarian or a farmer? Most people reply a librarian, even though there are twenty times more male farmers in America than librarians.

Despite all of that, Kahneman tells his audience that "the real hero of human thinking is system one, fast, automatic thinking." Fast thinking is what allows people to navigate daily life in real time, and most of the time it does very well. System-one thinking generates inferences and predictions. It works, Kahneman says, from "a certain computational model, a network of ideas and associations." A person's fast thinking looks back to the past for possible causes and interprets a likely future, one that is emotionally and socially coherent with that person's experience. "Most of what I think I know," Kahneman observes, "is what I've been told by people I trust and love." And robust system-one thinking, he says, is defined as "the maintenance and updating of a rich and accurate model of the world."

Given the "marvel of system-one thinking," Kahneman suggests that a very useful role for an "intelligent" computer system would be a kind of over-the-shoulder critic to assist human decision making. It could, he says, bring the slow-thinking perspective, with its careful parsing of evidence, to supplement and enrich fast thinking by humans. There are encouraging signs, Kahneman believes, that such quantitative assists can sharpen fast thinking. As evidence, he points to the recent work of Philip Tetlock, a professor of psychology and management at the University of Pennsylvania. Tetlock is best known for his 2005 book, *Expert Political Judgment: How Good Is It? How Can We Know?* Based on two decades of research, examining the predictions of hundreds of experts including professors, government officials, and journalists, Tetlock found that their long-term forecasts about political and world events were no better than chance. You might as well flip a coin.

But Tetlock's recent research is more optimistic about human prediction, showing how measurement and data can enhance human judgment. In a multiyear project sponsored by the government's Intelligence Advanced Research Projects Activities (IARPA), thousands of forecasters are making predictions on hundreds of questions over time, and their accuracy is being tracked. They are asked to put a number on their prediction, a probability (0 = no chance; 1 = sure thing). They are put in groups and given constant updates on their colleagues' forecasts on matters like exchange rates, the debt ratings of nations, and election results. Using this quantified, crowd-sourced approach, the forecasters in general have proved reasonably accurate in making predictions within shorter-term time spans, up to a year or so. And some have proved remarkably adept at refining their predictions, digesting other viewpoints and data, to recognize and overcome their biases. Tetlock calls them "super forecasters."

A clever helper. That is the benevolent portrayal of big data, as a tireless digital assistant to see what its overworked human boss missed. Dr. Herbert Chase, a professor of clinical medicine at Columbia University's College of Physicians and Surgeons, explains the potential of the technology with a story. Three decades ago, when he was young resident at Mount Sinai medical center in New York, a thirty-two-year-old woman came to the hospital with muscle weakness so severe that she couldn't walk or care for her young child. She received an extensive battery of tests, Chase recalls, that even was known as "the million-dollar workup." After the initial visit, Chase and his colleagues at Mount Sinai had no idea what was causing her muscle weakness. "The one smart thing I did was tell her to come back in a month," Chase says.

At the time, Chase was studying for his medical certification exam and had just about finished reading *Harrison's Principles of Internal Medicine*, a textbook of more than two thousand pages. He was going to skip the forty pages he hadn't read, but his wife shamed him into finishing. That section included a discussion of rickets, the softening and weakening of the bones, typically found in children who are severely malnourished or starved. Adult rickets is rare. The symptoms include low levels of phosphorus in the blood and muscle weakness—both of which the thirty-two-year-old woman had. "The lightbulb goes on as I read that," Chase recalls. As a teenager, the woman had been given a regimen of powerful antibiotics that had the unrecognized side effect of badly damaging her body's ability to absorb phosphorus, essential for storing and transporting energy in the cells. She was treated with doses of vitamin D and a phosphorus-rich diet, especially dairy products. Before long, she was walking and active.

Chase's description sounds like an episode of *House*, the television series about a curmudgeonly, intuitive medical genius, Gregory House, who miraculously comes up with diagnoses to baffling afflictions. "It was a triumph," Chase recalls. "But look at what had to happen. You had to get lucky." A slender physician with a graying beard, Chase has been a physician, a researcher, and an associate dean of education at the Yale School of Medicine before coming to Columbia to focus on biomedical informatics. He was an unpaid adviser to IBM's Watson project for health care applications, including developing Watson as a diagnostic assistant. When asked to try out the technology, Chase recalled the case so many years ago, with its rare diagnosis, as one sure to stump the software program. He fed in three symptoms, and Watson, which operates in probabilities, returned adult rickets as its second-ranked result. (Watson's first choice was hyperpara-thyroidism, overproduction of parathyroid hormone. "Not correct in this case, but plausible," Chase notes.)

For Chase, Watson's performance was a revelation, and a sign-post to what technology can do for medicine. Watson delivered a list of several potential diagnoses, with declining probability rat-ings. "We humans stop thinking after the third option," Chase says, while the machine offers far more alternatives to consider. The software, he says, looks promising and the stakes are high. Diagnosis is the pivotal starting point of treatment, determining the path taken, and an estimated 10 to 15 percent of all diagno-ses are mistaken, and many more are incomplete, accounting for as much as one third of all medical errors. Chase believes that a Watson-like technology will be part of the future of medicine, working in the background to assist doctors.

To appreciate the potential payoff, Chase says, it is necessary to acknowledge that the current practice of medicine is falling further behind in the information race. New research, evidence, and treatment guidelines are being generated at a torrid pace. Several hundred thousand new articles a year are added to the government's National Library of Medicine, whose collection of medical books, journals, and manuscripts numbers nineteen billion. No human can keep up. Watson, however, can scan and mine 100,000 medical journal articles in a few seconds. "We need a tool to pool and come up with the right evidence," Chase says.

Watson's future in medicine is far from certain. Even Dr. Martin Kohn, chief medical scientist for care delivery systems at IBM research, when we spoke in the fall of 2013, admitted that it will take time for Watson-style technology to really make in-roads in health care. Still, Kohn predicts that within five to ten years "Watson or something similar" will have made its way into primary care in America and elsewhere. Many physicians, Kohn says, "see real potential here, because everyone is feeling over-whelmed by information and would like help. . . . We make it clear Watson doesn't make decisions. But this technology can be a tool to make us make evidence-based decisions."

Kohn, who joined a big-data health start-up in 2014, hap-pened to be talking with IBM's product in mind. But Watson, though a pacesetter, is only one entrant in a burgeoning sector of technology intended to find and deliver timely, relevant data to people as they make decisions. The appeal is straightforward and irrefutable. Take the simple example that Dan Kahneman used of the man named "Steve" who was described as "meek," and people are asked to choose whether he is more likely a librarian or a farmer. The librarian choice is entirely logical, until you get the data Kahneman knew in advance. The data makes you rethink your intuition.

The prospect of such data-animated nudges to sharpen decision making, repeated countless times, up and down corporations, throughout the economy, is the why Erik Brynjolfsson believes big data will bring a "management revolution." Brynjolfsson is an economist at the Massachusetts Institute of Technology's Sloan School of Management, director of the MIT Center for Digi-tal Business, and an intellectual champion for the transformative power of big data. He is tall, with deep-set eyes, reddish-brown hair, and a trimmed beard and mustache. When he is talking excitedly about a subject, his voice occasionally cracks into a high-pitched range. These days, Brynjolfsson has a corner office with a view of the Charles River at the Sloan School, which occupies an austere, modern building of glass, steel, and limestone. His undergraduate major was applied maths at Harvard, his master's degree was in maths and decision science, and his PhD at MIT was in managerial economics. His research has focused on the impact of technology on productivity, work, and decision making.

I first started talking to Brynjolfsson in the mid-1990s. The Internet boom was under way, but there was a puzzle, economically. The Internet was minting millionaires aplenty and hailed as a revolutionary technology, yet it was not boosting the productivity of the American economy. Productivity gains—more wealth created per hour of labor—are the fuel of rising living standards, and a by-product of the efficiency that technology is supposed to generate. The conundrum raised the question of whether all of the investment in, and enthusiasm for, digital technology was justified. Robert Solow, a Nobel Prize–winning economist, tartly summed up the quandary in the late 1980s, when he wrote, "You can see the computer age everywhere but in the productivity statistics." Solow's critique became known as the productivity paradox.

Brynjolfsson, a technology optimist, has two answers for the skeptics. First, he argues, the official statistics do not fully capture the benefits of digital innovation. And second, he says that in technology, revolutions take time. To explain, Brynjolfsson points to his own work on technology and work practices, and to the research of others including a classic study by Paul David, an economic historian at Stanford. In his 1990 paper, "The Dynamo and the Computer: An Historical Perspective on the Modern Productivity Paradox," David observed that the electric motor was introduced in the early 1880s but did not generate discernible productivity gains until the 1920s. It took that long, he wrote, not only for the technology to be widely diffused but also for businesses to reorganize work around the industrial production line, the efficiency breakthrough of its day. By the 1990s, the adaptive response to the Internet was faster than with the electric motor, and by the late 1990s productivity rose.

In an immense economy, like that of the United States, with

its gross domestic product of $17 trillion, an amalgam of factors affects performance, including business cycles, financial crises, and demographic trends, not just technology. But Brynjolfsson sees a pattern playing out with big data that is comparable to past technologies. Innovations that have been percolating for years in research labs are making their way into products. An industry or two leads the way, like online advertising, and showcase projects point toward the future, like IBM's Watson or Google's self-driving cars (robotic incarnations of big data). Enthusiasm fans investment by companies and start-ups. But a broad-based payoff has not yet emerged. Debate rages between the techno-optimists and the pessimists.

In his office, I ask Brynjolfsson to describe the steps that led him to become a big-data believer. He starts by observing that the data groundwork has been laid in the steady digitization of business in recent years. In major corporations, he estimates that 90 percent of what companies do today, from communications to marketing to manufacturing, is either created in digital form (like e-mail or documents) or tracked digitally (like bar codes and radio tags). In the 1990s, Brynjolfsson figures that only about 20 percent of corporate activity left a digital footprint. "This explosion of digitization is a quantum change," he says.

M uch of Brynjolfsson's research has been looking inside companies. A research study that shaped his thinking, he says, involved examining the flow of e-mails inside an executive recruiting firm. Brynjolfsson and two graduate students looked at 125,000 e-mail messages, surveyed the recruiters, and collected the firm's accounting data on the recruiters' performance on 1,300 assignments over five years. A first cut of the data found that e-mail traffic did show patterns, like hours worked. "It was somewhat predictive," Brynjolfsson recalls, "but I was kind of disappointed. It was an obvious correlation."

But then his graduate-student researchers suggested analyzing the e-mail and other data from a social-network perspective—how widely and how quickly ideas spread, and identifying the influential communicators, the human hubs of information transmission. What they found was that people who were more connected got information faster, and they were typically the most successful recruiters. The study, first published in 2006, "Information, Technology and Information Worker Productivity," by Brynjolfsson and two coauthors, Sinan Aral, now a professor at MIT, and Marshall Van Alstyne, now a professor at Boston University, echoes work done three decades earlier by Thomas Allen.

In the 1970s, Allen, also an MIT professor, did his research in the physical world. He studied communication among engineers and the impact of the distance between their offices. He found out that the greater the distance between their offices, the less communication of technical information occurred—a distance-related falloff that could be quantified, known as the "Allen curve." Allen also identified the role of key communicators, which he termed "information gatekeepers." (In 2006, Allen, now a professor emeritus, revisited his research, and it reiterated the power of face-to-face communication and thus proximity. The more people saw each other in person, the more frequent their communications of all kinds, including by phone and e-mail.) The distinctive power of digital measurement is in both its scale and its detail. In the recruiter research, it meant being able to monitor communications down to words and phrases in 125,000 e-mails. "That was like the microscope, being able to see inside a corporation in a way you couldn't before," Brynjolfsson explains. "It's not just ten times better. It's data that is several orders of magnitude more fine-grained. That was a real aha moment for me."

Decisions fortified by data should be better decisions. That is intuitively obvious, and supporting anecdotal evidence keeps piling up. But Brynjolfsson broke new ground with research that measured the impact of the quantitative approach, not on a single company but broadly. Brynjolfsson and two other investigators, Lorin Hitt, a professor at the Wharton School of the University of Pennsylvania, and Heekyung Kim, a graduate student at MIT, collected detailed survey data from 179 large companies. In the surveys and follow-up interviews, they asked the companies about their business practices and investment on and use of technology. They asked not only how much and what kinds of data the companies gathered, but also how it was used—or not—in making crucial decisions, like whether to create a new product or service. They found that companies that had adopted "data-driven decision making" achieved productivity that was 5 to 6 percent higher than could be explained by other factors, including how much the companies invested in technology.

When I wrote a column for the *New York Times* in 2011 about this research, I asked Brynjolfsson what the real distinction is between a decision that is data-driven and one that is not. He replied with a crisp definition, and it was the first time I had heard it. The essential dividing line, he explained, is between decisions based mainly on "data and analysis" and on the traditional management arts of "experience and intuition."

In his office in late 2013, I make the case for the importance of experience and intuition, and the limits of the big-data mind-set. Brynjolfsson concedes that there is nothing sacred about data itself. "It's still possible to make a huge mistake with either approach," he says. "But I would argue that people are making a lot more of the experience-and-intuition mistakes than the data-and-analysis mistakes." The excesses of data-ism, Brynjolfsson suggests, are mainly worries for the future. "We have a long way to go yet," he says. "And the technology is well ahead of the management culture"—tracing a familiar historical pattern, first comes the innovation and then the organizational adaptation, from the electric motor to the Internet.

To get a glimpse of a company building a data-driven culture, the Denihan Hospitality Group is not at first glance an obvious candidate. It is a half-century-old, midsize hotel chain with fourteen hotels. It owns eight and manages them all, mostly in New York, with one each in Chicago, Miami, and Washington, DC; its brands include the Affinia hotels, and one-of-a-kind properties like The Surrey and The Benjamin. But the family-owned Denihan is an example of a conventional company that has made real progress toward using technology to make more of its decisions aided by data. And the hotel company has been at it seriously for more than a decade. Indeed, Denihan was cited as one of a few examples in a 2009 book, *Profiles in Performance: Business Intelligence Journeys and the Roadmap for Change*, by Howard Dresner, a respected business and technology consultant.

The Denihan experience combines leadership at the top, focused goals, and close cooperation between a small data team and the people managing hotels day in and day out. The investment wasn't huge, and the progress has come in measured steps. The current data program at Denihan had its origins in the late 1990s. Brooke Barrett, the co–chief executive, says she always took a "start with the facts" approach to managing the company. Yet by the late 1990s, she recalls, it was clear that the company wasn't efficiently handling the flood of data it was producing. An outside consultant was brought in and it found that Denihan had a lot of data, but little of it was being gathered, analyzed, and linked to decision-making activity in a coherent way. Excel spreadsheets alone, the consultant study concluded, were no longer enough.

To tackle the problem, Denihan hired its first chief information officer, John Cahill, recruited in 1999 from InterContinental Hotels, where he was a senior technology executive. Less than two years later, Cahill brought in Menka Uttamchandani, a database marketing specialist working for the Hyatt chain in Hong Kong. A crisp, efficient, fast-talking Indian, Uttamchandani got the assignment to set up a modern data-reporting and analysis capability. The goal, she says, has always been to use data as a tool for making decisions. She started alone, and today her business-intelligence unit has only four people. Their work, she explains, is guided by three objectives: to boost revenues, curb costs, and guide the company's strategic direction.

Uttamchandani's team can parse data in hundreds of ways to examine guest bookings by geography, demographics, sales channel, and many other variables. Once at a hotel, guest spending patterns are tracked by type—room, spa, restaurant, telecommunications, entertainment. The median booking is thirty-five days in advance. But with Denihan's business split about fifty-fifty between business and leisure travel, the variation is considerable, from rooms booked several months in advance to rooms booked the same day. The data analysts can sift corporate customers by type and booking history. So, for example, the system can flag a customer who booked rooms last year for an annual industry conference but not yet for this year's conference. Total value calculations by customer are possible; for example, a corporate customer that produces $10 million a year may not be as profitable as one that books $6 million, if the latter is generating business in slow times, when rooms might otherwise be vacant. The effectiveness of incentives can be tested: Does the offer of free parking or a free breakfast bring in more responses? Real-world science experiments in marketing become possible.

The many ways data can be sliced and diced is important.

But Uttamchandani has also worked closely with Denihan's executives, hotel managers, booking strategists (called "revenue managers"), sales managers, and marketing managers. Those are her customers, and she has catered to them with daily reports showing the pulse of the business, across several dimensions. Her group has developed simplified dashboards, tailored both for different kinds of managers and for different devices—smartphones or desktop computers. The design work to make the software easier to use and more helpful has fostered adoption. "If four or five people routinely used Menka's business-intelligence tools when she started, that was a miracle," Barrett recalls. "Now, all the executives and management staff in the hotels do," about 150 people.

There was skepticism at first. Brian Gehlich is the general manager of the Affinia 50 hotel, at Fiftieth Street on Manhattan's East Side, just off Third Avenue. He has worked in the hotel business for nearly three decades and has spent the last twenty years with Denihan. When Uttamchandani came aboard, promising that her technology would be helpful and user-friendly, Gehlich recalls thinking, "Yeah sure." So much of business technology, he says, takes time to master and adds work for frontline managers. It is technology used as an instrument of command and control from above. But the data tools, he insists, have proved to be a different category of technology, a servant rather than a master. They deliver timely information *to* him. "Menka and her team really can translate the data and make it understandable and useful," he says.

Gehlich is a convert. Early each morning, he checks his smartphone for the daily "pace" report that shows occupancy, average room rate, and new bookings and cancellations for the previous day. When I visit him one morning in December 2013, Gehlich is in a small office at the hotel. Behind his desk chair are two guitars. Guests wanted to play in their room the night before, and Gehlich rented a pair of guitars. "It's a personal service business," he observes.

On his desktop computer, Gehlich is looking at a data graph that tracks the trend of future business out three months. Bookings from the major corporate travel agencies, like American Express and Carlson, are trending down compared with last year. He clicks through to dig deeper and finds that a couple of promotions—10 percent off for some days, and free breakfasts for others—ran last year that are not being offered again this year. "Did someone make that decision? Should we rethink it?" he says. "It's key stuff."

For Gehlich, the business-trend tool is an early-warning system—much like Dr. Buchman's software at the Emory medical center for spotting intensive care patients whose biological data signals show them heading for trouble. With large corporate accounts, Denihan's hotels, like others, negotiate deals—discounted rates for guaranteed volumes of business.

Today, Denihan's hotel managers go into those talks with data that shows how profitable a corporate customer is, depending on when it books rooms, how many, and what kind. A few months earlier, Gehlich notes, he decided not to renew a deal with a major bank because it would not have been profitable. "Without that bottom-line data on each account, I never would have done that," Gehlich says. And Denihan's ability to see profit-and-loss data down to the level of individual customers mirrors what McKesson can now do, on a far larger scale, with the customers in its nationwide drug distribution network.

I ask Gehlich about the payoff from Denihan's data strategy. He replies that it is "enormously valuable" and estimates that the improvement in terms of revenue and productivity is "well into double-digit range—it has to be." But mainly, Gehlich talks about how much more of his business is measured and how that has changed the way decisions are made. To be sure, most hotel chains have moved to adopt digital technology over the last decade or so. Before, hotels relied on paper, spreadsheets, and registration cards that guests filled out when they checked in. Much of that has been automated with computerized "property management systems" to capture information at the front desk and elsewhere. And nearly all hotel chains now subscribe to data services that report the prices and pricing trends at competing hotels, with most of the data scraped off the hotels' websites.

But the detailed data analysis done at Denihan, which is widely available to managers, is unusual. The data tools, Gehlich says, have become a vehicle for managing his hotel with a different mind-set. "This company decided years ago that this is the future—better data used intelligently to make better decisions," he explains.

Barrett, the co–chief executive, offers the same kind of answer when I ask her about the measureable benefit of the company's data program. Sure, she says, she can point to specific data-driven projects. One was a win-back project for identifying "lapsed guests," who used to come regularly to Denihan hotels but stopped, and for offering them incentives to return. The incentives are calibrated, based on the guests' spending history at a Denihan hotel. The return on investment was 300 percent for the win-back initiative, and what began as a one-off project is now a continuing practice.

But Barrett concedes that Denihan has never done a cost-benefit analysis of the investment in Uttamchandani's business-intelligence group. Her reply, in essence, is that the *qualitative* difference in management decision making has been so apparent that doing a hard *quantitative* assessment would be unnecessary— and not capture the sort of benefits that are hard to pin a number on. "It encourages people to ask all sorts of what-if questions and see things differently," Barrett says. "As a mentality for making better, more grounded decisions, it has been critical. And it's been a journey."

Today Denihan is going further, collaborating both with IBM and a start-up, Duetto Research, on advanced data projects. To help with this next step, Denihan has recruited Thomas Botts, a former executive at Starwood Hotels and Delta Air Lines, as its chief customer officer. The objective is to use big data to generate fine-grained, accurate pricing predictions, make personalized offers on the fly, and react in real time to changes in the marketplace.

The smart software works like this: A hotel revenue manager might get an alert on her smartphone that bookings on the hotel's website for a certain day, several months in the future, are increasing sharply. The software recommends raising the room rate from $190 to $225. The manager clicks through to see a report that in the past similar activity on the website has been associated with a rival hotel having excess demand or a local event being announced. She sees the software's rationale, agrees, and clicks to approve the price increase. The change instantly goes into effect on the hotel's website and reservations system.

In the future, personalized marketing offers might be delivered to hotel guests' smartphones, with their approval. The hotel's software notices that it's a slow afternoon at the bar and, as you walk by, it sends you an offer for a free drink or half off. "We're working on it," Botts says.

5

THE RISE OF THE
DATA SCIENTIST

The San Francisco office of Cloudera resides on the eleventh floor of a modern office building on California Street in the city's financial district. In recent years, a San Francisco office for fast-growing technology companies headquartered in Silicon Valley has become a popular amenity—and one that reflects the shifting topography of entrepreneurialism, talent, and taste in northern California's Bay Area.

Not so long ago, the boundaries were clear-cut. Silicon Valley had all of the start-ups and the engineering wizardry. San Francisco had trendy culture, fine dining, and old money. But the Valley's monopoly on entrepreneurial vigor has been broken decisively by the rise of San Francisco–based technology companies ranging from Twitter to Salesforce.com, a supplier of Internet software for companies. So the terms of trade have been altered for the region's young high-tech workforce. Many of them view the Valley as a great place to work, but no match for San Francisco as a place to live. To accommodate them, companies founded in the Valley often have sizable offices in San Francisco, like Cloudera's.

The wall behind the reception desk is adorned with an abstract mural, with countless strings of letters and bits (1's and 0's) in spiraling cascades, suggesting a kind of digital helix. In front of it, in silver metal letters, is the company name, Cloudera. Below, in blue letters, is the tagline: Ask Bigger Questions.

When a computer switches between one task and another, it is called toggling. Jeff Hammerbacher toggles between informality and intensity. He can be casual and unguarded in conversation, but you sense that intensity is his natural state. Show up for a meeting, and he at first seems genuinely surprised to see you. His mind was elsewhere, and you showed up. "Oh, you're here." It is a crisp February afternoon in San Francisco. Hammerbacher's desk is piled high with books of every description; when he is gone, his colleagues at Cloudera treat the book pile as a lending library. Hammerbacher travels light—a MacBook Air and an iPad, and sometimes his Kindle, too, in a small backpack, usually with a couple of ink-on-paper books he's reading at the time.

C loudera was Hammerbacher's next act after Facebook. Bring up the subject of Facebook with him, and the conversation covers a lot of ground. Facebook was a veritable university of the data arts and sciences for Hammerbacher. He was there for less than three years, from early 2006 to late 2008. When he left, Hammerbacher says, he departed not so much to help found Cloudera but mainly to leave Facebook. By the time of his exit, Hammerbacher says, he had decided that "the mission of the company was not that motivating to me. Ultimately, I don't care about social networks. . . . To me, Facebook seems to make life more quotidian," trivializing and commercializing human communication. His Facebook page is long dormant.

Hammerbacher sees the ascendant consumer Internet companies—Google, Facebook, and Twitter—as flawed success stories. They make their money selling advertising, and Hammerbacher despairs that so much computer science brainpower is dedicated to targeting online ads. He expressed his reservations most pointedly in 2011, when he told Ashlee Vance of *Bloomberg Businessweek*, "The best minds of my generation are thinking about how to make people click ads. That sucks." It is an observation that has a ring of truth. But it is also true that new technologies always go first to where it is easiest to make money, and then spread more broadly. The early markets for the printing press, in addition to the Gutenberg Bible, were religious tracts, political screeds, and pornography. Only later did the printing press become a vehicle for democratizing knowledge and for mass education.

Online advertising is an economic virtue. It brings a new level of efficiency to the market for advertising, reducing costs, and freeing money and resources for investment elsewhere in the economy. To that, Hammerbacher offers a personal yet perfectly reasonable observation. "So why not work on the elsewhere?" he asks in reply. "Work on things that matter more to you."

Hammerbacher did just that, following his principles and walking away from a lot of money by leaving Facebook when he did. But his objection to the giants of social networking seems to be mainly a matter of taste—just as critics of television, and radio before, bemoaned what they regarded as a descent into crass commercialization and lowbrow culture. But Facebook and Twitter are also enormously valuable social utilities for personal communication, sharing information, political protest, and, yes, commerce. And they have advanced computing and data science. Facebook, in particular, has pushed forward the development of open-source software that other companies can use freely, enriching the common pool of big-data technology. If you make that case to Hammerbacher, he does not so much back down as step to the side, offering an olive branch. "I recognize all that, and that this is an incredibly nuanced argument," he says. "Facebook has had a huge influence on the world, and not to see both sides of that is dishonest."

The Facebook years were Hammerbacher's formative years as a data scientist, the source of his financial freedom, and, to some degree, his professional reputation as "the guy who built the data team at Facebook," even though he left more than six years ago. When he showed up at Facebook in early 2006, Hammerbacher was twenty-three, older than most. Mark Zuckerberg, founder and chief executive, was twenty-one. Zuckerberg, along with two early Facebook employees, Charlie Cheever and Dave Fetterman, were all acquaintances from Harvard. Another person Hammerbacher talked to before he signed on was Jeff Rothschild, who was fifty at the time. Rothschild had been sent over from the venture capital firm Accel Partners, a major investor in Facebook, to provide the proverbial adult supervision that so many start-ups require, and he stayed.

Hammerbacher liked the feel of the loose, scrappy start-up with its sky's-the-limit ambitions, which would prove accurate only in retrospect. The cluttered office in Palo Alto, inhabited mainly by twentysomethings dressed in jeans and T-shirts, was a world apart from the Wall Street trading floor he departed in New York. Intellectually, he was fascinated by the emerging research in online social-network analysis. He read published papers and books in that field, including the work of Duncan J. Watts, a network scientist and then a professor of sociology at Columbia, and even talked to him. (Watts is now a researcher at Microsoft Research.)

Online social networks have become a new laboratory for studying human behavior on a scale never done before, making social science far more quantitative. To appreciate the difference, look back to the 1960s, when Stanley Milgram of Harvard used packages as his research medium in a famous experiment in social connections. He sent packages to volunteers in the Midwest, instructing them to get the packages to strangers in Boston, but not directly; participants could mail a package only to someone they knew. The average number of times a package changed hands on its way to the final destination was remarkably few, about six. It was a classic demonstration of the "small-world phenomenon," captured in the popular phrase "six degrees of separation."

Today, social-network research involves mining huge data sets of collective behavior online, a digital fishbowl for observing human activity. Among the findings: people whom you know but don't communicate with often—"weak ties," in sociology—are the best sources of tips about job openings. They travel in slightly different social worlds than close friends, so they see opportunities you and your best friends do not.

Another insight, from research published in 2013, is that the shape of a person's social network is a powerful signal that can identify one's spouse or romantic partner—and even if a relationship is likely to break up. The scientists found, somewhat surprisingly, that the total number of mutual friends two people share—embeddedness, in social-networking terms—is actually a fairly weak indicator of romantic relationships. Far better, they found, was a network measure they call dispersion. This yardstick measures mutual friends, but also the farther-flung reaches of a person's network neighborhood. High dispersion occurs when a couple's mutual friends are not well connected with each other. A spouse or romantic partner, it turns out, is a bridge between a person's different social worlds. Facebook data, stripped of personally identifying information, was the lab for the research—1.3 million adults who listed a spouse or relationship partner in their online profile, with 8.6 billion links to others, tracked every two months for two years. When a couple scored low on the dispersion algorithm, they were 50 percent more likely to break up in the following two months. "This is the kind of thing you can only do with big data," explains Jon Kleinberg, a computer scientist at Cornell, and coauthor of the research paper, with Lars Backstrom, a senior engineer at Facebook.

When Hammerbacher arrived at the two-year-old start-up, he had no illusions about writing research papers. That would come later for Facebook. Hammerbacher's notion of social-network analysis was to use Facebook's data to improve its service. But he found that Facebook had "zero infrastructure" for doing that kind of data analysis. That was not so surprising for a young company with the growth trajectory of a rocket ship. The "infrastructure" that mattered most was the hardware and software needed to keep Facebook up and running smoothly. Hammerbacher spent his first six months writing software to pluck selected data on traffic, use patterns, and messages and store it in a separate database. "It was first just a matter of building this nest of data," he explains, "that we could use to start improving the product and thinking about how the business worked." He often had to make do with the software tools designed for the old data world—when data was comparatively sedate and small. In the old model, data fit into the neat rows and columns of traditional databases, and most data came from internal company reports and spreadsheets. But the Web is a data Wild West. It is the unruly data of Web pages, text messages, photos, and videos uploaded by users, and, even in 2006, it was accumulating at a torrid pace at Facebook. The rising data tide was a by-product of pleasing growth for the social network, but Hammerbacher saw it as a valuable asset that was going unused.

H ammerbacher's boss was Adam D'Angelo, the chief technology officer at Facebook. A young computer whiz, D'Angelo had excelled in national and international programming contests in high school and college, attending the California Institute of Technology. At Facebook he had the added advantage of knowing Mark Zuckerberg since high school, at Phillips Exeter Academy, an elite prep school. Hammerbacher urged D'Angelo to make the investment to build up a data analysis team and create the computing engine for the job. At the time, Facebook was moving beyond its origins as a service for university students. In September 2006, it opened up to anyone thirteen years old or older with a valid e-mail address. Facebook was no longer a social network created by college students for college students. The decision making would get trickier. "It was time to do things based on data," D'Angelo recalls. "It was Jeff's intuition, and I agreed with it. But it was a departure at the time."

Hammerbacher's first hire was Itamar Rosenn, a graduate student at Stanford at the time. He shared an apartment in Palo Alto with two early Facebook employees. They seemed to be having more fun than he was in graduate school, and Facebook was hiring. He and Hammerbacher first met at a local Chinese restaurant, and Rosenn had the kind of background and interests that would intrigue Hammerbacher and make for a long evening. As an undergraduate at Stanford, Rosenn had combined the study of computers and human cognition with the university's version of a great books program, called structured liberal education. His graduate study focused on economics and econometrics—its appeal being in "applying quantitative thinking to social problems," as Rosenn put it. Facebook wasn't really such a stretch.

Rosenn recalls that first dinner meeting with Hammerbacher as "a sprawling conversation" that covered topics including artificial intelligence, hypothesis testing, econometrics, statistical inference, and the social sciences. When it ended, Hammerbacher gave him a "homework assignment" to sketch out a system for sending e-mail invitations to join Facebook that would maximize growth. Rosenn apparently passed the test and went to work for Facebook at the start of 2007, becoming the first member of Hammerbacher's data team, which would grow to about thirty people by the time he left in the fall of 2008.

Rosenn remains at Facebook, and his job title is manager of measurement systems. Of course, the data team at Facebook can do far more now than in the early days; the team is bigger, and so is the data, and the software tools are better. But even today, Rosenn observes that "much of the time you are just doing counting." There are just so many measurements that can be made with the data, sliced and scrutinized in so many different ways—users' behavior on Facebook by click streams, messages posted, news stories shared, photos uploaded, and others, according to their demographics, geography, employers, schools, native language, and "vintage" (when they joined Facebook), in potentially endless combinations.

In his first years, Rosenn worked on preparing the "growth report" every two weeks. With data and graphs, the growth report tracked where and with whom Facebook was propagating like digital kudzu and where it was not catching on. The growth report guided operations and strategy, where to put more investment and manpower. It was an important number to count, but by no means the only one, and that was especially true as Facebook got a little older. Another vital metric was what in the Web world is often called "engagement," which is typically measured by the time a user spends on a website (in minutes a month, for example). Engagement times, along with page views and ad impressions, are conventional measures of a website's success. Yet Facebook took a broader view. What it measured very closely was the overall flow of information among its members—posts, status updates, news articles sent to friends, or any acts of creating and sharing information rather than just being consumers of information. Page views and ad impressions sometimes were sacrificed if a decision, like the design of user pages, was likely to increase information sharing.

C ounting is political, social scientists say. What they mean is that the selection of what to count reflects the values and biases of the people doing the counting. And it is as true in business as it is in politics. His choice of words is different, but Jeff Rothschild, over a breakfast in Palo Alto, made the same kind of measurement-and-values connection, when explaining Facebook's success. Rothschild is a tall man, with neatly trimmed gray hair and a wealth of experience in the technology business. He was a cofounder of Veritas, a maker of data storage management software. Symantec, a larger software company, bought Veritas for $13.5 billion. At Facebook, Rothschild was vice president of infrastructure engineering—the guy in charge of keeping the site humming—for five years, until 2010. Today, he works as a venture capitalist at Accel, and still spends one day a week at Facebook, a part-time elder statesman, advising on projects and mentoring young engineers.

Facebook became a major company in part because it pursued a larger goal, Rothschild explains. "It optimized," he says, "for the information flow that is traversing the arc" of its network. Rothschild has spent most of his career in the data business; his training may be as an engineer, yet Rothschild has thought a lot not only about how to manage data but also how to manage *with* data—the real business opportunity afforded by big data. "It's not about the size of the data," he says. "It's about being able to collect it and then steer the organization based on the metrics you really most value in the long run."

Hammerbacher, Rothschild recalls, was a forceful, intense, and effective team leader, if at times distracted. He had "the confidence of an entrepreneur," Rothschild adds. "He assumed he could get it done, and did get it done." What was done was the building and refining of software tools for collecting, indexing, and asking questions of massive volumes of Facebook data. Speed was essential. With one software system, a day's worth of clickstream data took twenty-four hours to process. An improved program did the job in less than two hours. But their early software for handling Web data was good at only one task.

What was needed was a more general-purpose tool, tailored for handling big data and that could become the software foundation on which other programs run. In computing, such tools are called "platforms." The complementary programs add new capabilities and diversity to a flourishing digital ecosystem, increasing the value and popularity of the host platform and the software applications that run on it. The best-known technology platforms are operating systems. In the personal computer industry, Microsoft's Windows operating system became the dominant platform. In smartphones, Google's Android operating system is the leader in market share. But Apple's iOS software, the pioneer platform in smartphones, is a strong rival with a large and loyal following among application developers, not to mention smartphone owners. In data centers, the Linux operating system is a popular platform for data-serving computers.

Linux and Android are both open-source software, distributed free with its underlying "source" code published, unlike the proprietary offerings from Microsoft and Apple, for example. Programmers can tweak and modify open-source programs within certain rules. In late 2005, a potential big-data platform emerged, called Hadoop, an open-source project, begun by two engineers, Mike Cafarella and Doug Cutting. The quirky name was what Cutting's toddler son called his favorite stuffed toy, a yellow elephant. Hadoop is an open-source variant of Google's technology for splitting up and then processing large data sets across many computers. Hadoop was promising, but it was slow at first. In the spring of 2006, Hadoop could sort through about 2 terabytes of data in roughly two days. Two years later, Hadoop could do the same job in a few minutes. The Web portal Yahoo!, which hired Cutting, made the early investment of people and resources to improve Hadoop. (Today, there are specialist companies that distribute and provide technical support for Hadoop, led by Cloudera, where Hammerbacher is chief scientist and Cutting is chief architect.)

Seeing the rapid pace of improvement, Hammerbacher championed the adoption of Hadoop at Facebook. He convinced the company to set up its first sizable cluster of computers running Hadoop in early 2008, and things took off from there. Facebook would make significant contributions to the Hadoop open-source code during Hammerbacher's tenure and long after. Under Hammerbacher, for example, Facebook began the development of a program known as Hive, which runs on top of Hadoop. Hive is a step toward data democratization, a simplified software tool, so that you don't have to be a computer scientist to ask questions of the data. Like Hadoop itself, Hive is freely available under an unrestrictive open-source license, called Apache. The names may be quirky and geeky, but these software building blocks matter a lot. They are steps toward bringing big-data technology into the mainstream, so that data exploration is not confined to an aristocracy of experts.

The history of computing is the story of technology being democratized. In the 1960s, mainframe computers were powerful engines of technical progress, but they were confined to big companies, major universities, and government agencies. In the

1970s, minicomputers opened computing up to more people in business and academia. But it was not until the 1980s that the personal computer revolution brought computing to the masses, with both low-cost hardware and low-cost software that nontechnical people could use. In business, the classic example was the spreadsheet, which opened up financial analysis to anyone with a PC. Today, big data is in the equivalent of its mainframe era, or perhaps its minicomputer phase. But start-ups, big companies, and open-source projects in Silicon Valley and elsewhere are hard at work on the big-data spreadsheet—typically a Web-based dashboard that taps into the data beneath.

There seems little question that Hammerbacher left a legacy of accomplishment in his few years at Facebook. He was an advocate for open-source software and he encouraged the members of his team to attend scientific conferences and publish papers, sharing what they learned. He also started the study of data and data experiments to guide business decisions. Such experiments, in the Web world, are known as A/B testing. Mostly, these are simple randomized tests of what works best. For example, a designer might come up with a new page layout that changes the location of an icon for status updates, adding photos or some other Facebook feature. One group of Facebook users would see the site without the change (group A), while another demographically similar set of users (group B) would be presented with the proposed design change. A/B testing is routine now in website development, online advertising, and marketing. Yet before Hammerbacher's data team, there was no A/B testing at Facebook. It would all become far more sophisticated in the years after Hammerbacher left. "But he laid the foundations for doing data analysis at scale," Rosenn acknowledges.

The conscripts to Hammerbacher's data team were at first given one of two job titles. Some were data analysts, and others were research scientists. The division was less a job description than a nod to academic lineage. If you had a PhD, you were a research scientist. Hammerbacher was an exception, a research scientist despite his degree deficiency. But the distinction seemed artificial, as the work they did was increasingly an amalgam of skills, combining computer science, business, and social science. So, Hammerbacher says, "We decided to mush those two titles together and call them data scientists." At first, a few PhDs resisted, viewing the change as a loss of title prestige. "But ultimately everyone embraced it, and it took on a life of its own," he observes. And to him, it seemed natural. "Data science is what we did."

The origins of data science reach back half a century or more. Hammerbacher's choice of terms wasn't mere happenstance. As soon as he accepted the job at Facebook, Hammerbacher began poring through technical papers and books that provided clues to the evolution of data science. In the spring of 2012, he taught a course in data science at the University of California at Berkeley. His first talk included a brisk yet comprehensive tour of the pertinent literature. His survey stretched from the first half of the twentieth century and the English statistician and biologist Ronald A. Fisher, who did pioneering work on the design of experiments using agricultural and gene data; to Hans Peter Luhn, the IBM scientist whose paper in the late 1950s imagined a computerized "business intelligence" system for mining information to improve business decisions; to contemporary computer scientists who have authored important works on data and discovery— Jim Gray, Tom Mitchell, Randy Bryant, and Peter Norvig. For each, Hammerbacher cited that person's contribution, and what he had learned from reading each one's books or papers. It was part bibliography and part personal tutorial.

In the Berkeley lecture, Hammerbacher singled out John Tukey as "the first data scientist." Tukey, a researcher at Bell Labs and a Princeton professor, who died in 2000, was a multitalented mathematician. In World War II, his maths models were used to improve the accuracy of artillery firing and bombing drops. He was also a frequent consultant to business and government, applying statistical techniques to improve the accuracy of tasks from collecting census data to predicting election-night presidential contests for television. Tukey was early to recognize the importance of computers and the role they could play in transforming statistics and data analysis. And he was something of a wordsmith. He is credited with coining the term "bit," a contraction of "binary digit," the lingua franca of computing. And the first published use of the word "software," as a computing term, was in an article in 1958, written by Tukey.

In 1962, Tukey wrote an influential paper, published in the *Annals of Mathematical Statistics*, titled "The Future of Data Analysis." He began, "For a long time I have thought I was a statistician, interested in inferences from the particular to the general." But as his work and the use of maths and statistics evolved, he added, "I have had cause to wonder and to doubt." And a few sentences later, he wrote, "All in all, I have come to feel that my central interest is in data analysis." To Hammerbacher, Tukey was the founding father of the "data-first" ethos, a kind of Copernican shift in discovery and decision making. The data-first proponents, he explains, are "starting with the data and seeing what it tells them instead of starting with a hypothesis and seeing what they can learn."

Tukey himself didn't use the term "data science." And today, there are debates and differing definitions about precisely what data science is. That isn't surprising. Uncertainty and experimentation while pursuing a new set of problems and opportunities are how disciplines emerge in technology. In the postwar years, big computers were the disruptive technology of the day, with the potential to transform scientific research, business, and government operations. To really create a computer age, skilled people and new tools and techniques were needed. In the 1960s, universities responded with programs in computer science, a new discipline that combined mathematics and electrical engineering.

We see a similar pattern with data science. It is certainly where established academic departments, like statistics and computer science, are headed, and have been for a while. Back in 2001, William S. Cleveland, then a researcher at Bell Labs, wrote a paper he called an "action plan" for essentially redefining statistics as an engineering task. "The altered field," he wrote, "will be called 'data science.'" In his paper, Cleveland, who is now a professor of statistics and computer science at Purdue University, described the contours of this new field. Data science, he said, would touch all disciplines of study and require the development of new statistical models, new computing tools, and educational programs in schools and corporations. Cleveland's vision of a new field is now rapidly gaining momentum. The federal government, universities, and foundations are funding data science initiatives. Nearly all of these efforts are multidisciplinary melting pots that seek to bring together teams of computer scientists, statisticians, and mathematicians with experts who bring piles of data and unanswered questions from biology, astronomy, business and finance, public health, and elsewhere.

Data science may or may not become its own academic discipline in the traditional sense—a department or college of data science. Data science could instead merely become ubiquitous, as a cluster of skills and ways of thinking employed in every field of inquiry. Edward Lazowska, a computer scientist at the University of Washington, believes that data and data science will be a common ground and common language across disciplines—"the great unifier of the next decade or two," as he puts it. However the field develops, data science talent is going to be in demand, and universities are struggling to compete with industry. In 2013, three universities—the University of Washington, New York University, and the University of California at Berkeley—won a $38 million grant jointly from the Moore and Sloan foundations to help foster a "culture of data science" in their institutions. Creating high-status career paths is a vital ingredient in the recipe. In their proposal to the foundations, the project leaders of the three universities wrote, "We need them back at the universities, working on the world's most important science problems—not trying to make people click on ads."

When Hammerbacher left Facebook in the fall of 2008, the timing was the only thing that surprised Jeff Rothschild. Just months earlier, Rothschild had introduced Hammerbacher to his partners at the venture capital firm, Accel Partners. Rothschild saw entrepreneurial potential in Hammerbacher, a young man with "a strong personality and very ambitious." He always assumed Hammerbacher would leave Facebook, but he thought it would be a few years later, not a few months after he had opened the door at Accel. "You can't do that," Rothschild told him at first. But Rothschild soon came around. "It shouldn't have surprised me," he says. "Jeff's not the kind of guy to wait around."

For Hammerbacher, the decision to leave Facebook, like his other career moves, was not about money. He is certainly wealthy, by any normal standard. One day, sitting on stools and having lunch at a Joy Burger in Manhattan, I ask Hammerbacher the money question. "You're not really rich, are you?" No, no, he replies, almost dismissively. Then, by way of explanation, he adds, "I own three houses and have ten million dollars in the bank." Yes, to most people that is astronomical affluence, especially for someone who was just thirty years old at the time. Yet Hammerbacher wasn't being facetious. After he left in 2008, he sold his shares in Facebook in the private marketplace for stock in companies before they go public. But Facebook was growing like a rocket. Had he stayed on until Facebook first sold shares in the public stock market in May 2012, Hammerbacher would have been much richer. He left Facebook anyway, knowing he was taking only a small fraction of the financial windfall he could have had. His father, Glenn, recalls once asking Jeff whether he had second thoughts about walking away early and leaving so much money behind. "Not at all," Jeff replied. "I got a good price for my Facebook stock. And how much money does anyone need anyway? How much is enough?"

Being nonchalant about money, of course, is easier when you have plenty. And wealth gives Hammerbacher the freedom to do what he wants. Even greater wealth may well come from his stake as a cofounder of Cloudera. In April 2014, the chipmaker Intel and venture capital firms invested $900 million in Cloudera, valuing the young company at $4.1 billion. But money is not his principal motivation; it is not the measure that matters most. The same seems true of Hammerbacher's friends and colleagues who are technically adept. They are nearly all his age, plus or minus a couple of years, late twenties to early thirties, entrepreneurs, software designers, or data quants. One or two are very wealthy, but most are not, at least not yet. They evaluate work based on whether the job is exciting and important in some way. Hammerbacher himself applies a four-point criterion: "My personal interest, my personal capability, the needs of society, and the tractability of the problem." In short, work should be intellectual fun, be important, and provide a real opportunity to "get things done," as engineers so often put it.

Across much of the economy, industries are being upended left and right, often by the digital technologies people like Hammerbacher create—and good jobs are considered blessings, scarce and dear. But Hammerbacher and his peers reside in an alternative economic universe. They live without any real sense of economic risk. For them, there are plenty of Plan Bs waiting. If one thing doesn't work out—a start-up goes under or the appeal of a job fades—they can easily move on to the next. Their skills give them their aura of personal economic confidence that they will always make a good living, and perhaps a killing someday.

A t Cloudera, Philip Zeyliger was an early recruit. Hammerbacher calls Zeyliger one of the many "converted mathematicians wandering around in data space." A Harvard maths major, Zeyliger first went to work on Wall Street for D. E. Shaw, a large investment management firm that is known for its sophisticated use of quantitative computer models and for hiring Ivy League maths and computer science whizzes. He liked the company and the people there, but the work did not appeal to him. After two years, he jumped to Google as an engineer. The transition from Wall Street to the Internet giant wasn't difficult. "What we lovingly call a data scientist today is very similar to a quant," he says. "It's mathy, intelligent human being stuff."

In the fall of 2008, Hammerbacher invited Zeyliger to a Vietnamese restaurant in Palo Alto, to convince him to leave Google and join Cloudera, which was just being founded at the time. Over a noodle dinner, Hammerbacher explained the plan. The start-up was going to take Hadoop, the open-source big-data software, and bring that technology beyond its bailiwick in a handful of Internet companies like Google and Facebook. The four founders were three data engineers from the Web giants—Christophe Bisciglia of Google, Amr Awadallah of Yahoo!, and Hammerbacher—and Mike Olson, a former executive at Oracle, the largest supplier of corporate database software. Hammerbacher, Zeyliger recalls, portrayed the start-up as a force for digital democracy, bringing big-data power tools to the rest of the economy. He was convincing, and Zeyliger was excited. But Zeyliger had a reservation. He was recently married and concerned about the endless hours of the start-up life, or "work-life balance," as he put it. In reply, Hammerbacher offered a certain backhanded reassurance. "If you love what you do," he told Zeyliger, "you don't notice." Years later, Zeyliger, sitting in Cloudera's San Francisco office and smiling, says, "He was pretty much right about that." Zeyliger was employee No. 8 at Cloudera, which now has more than 400 employees.

At Cloudera, Hammerbacher still keeps a hand in, recruiting engineers and meeting customers, but he isn't involved in the day-to-day operations of the company. But he does take time away from Mount Sinai for Cloudera business.

Hammerbacher is particularly convincing and authoritative, in making the case for next-generation data technology to potential customers.

Since he is good at explaining the promise of data, Hammerbacher has been frequently asked to do it, at conferences and in meetings. He does it selectively, but one speaking engagement in 2010 focused his interest and steered his career in a new direction. He had agreed to give a talk in Seattle at a conference hosted by Sage Bionetworks, a nonprofit organization dedicated to accelerate the sharing of data for biological research. Hammerbacher knew the two medical researchers who had founded the nonprofit, Stephen Friend and Eric Schadt. He had talked to them about how they might use big-data software to cope with the data explosion in bioinformatics and genomics. But the preparation for the speech forced him to really think about biology and technology, reading up and talking to people.

The more Hammerbacher looked into it, the more intriguing the subject looked. Biological research, he says, could go the way of finance with its closed, proprietary systems and data being hoarded rather than shared. Or, he says, it could "go the way of the Web"—that is, toward openness. The goal, he adds, should be a "scientific commons for disease modeling and drug discovery." And the field seemed fairly open and in need of the software expertise he could bring. Medicine ranked high in three of the categories on his checklist for deciding how to spend his time: his interest, his capability, and society's needs. In health care, the stakes could not be higher; people's quality of living and their very lives are the outcomes. The question was how much good data skills could do. How much difference could they make in medicine? A big question, but Hammerbacher found the challenge appealing, and he remembers thinking to himself, "This is the best problem."

6

DATA STORYTELLING:
CORRELATION AND CONTEXT

Y ou're given one data point: 39. What does it tell you? Not much. It's a number greater than 38 and less than 40. Anything beyond that would be speculation and conjecture. Next, you're presented an additional piece of information, 39 degrees. It could be the measure of an angle or a temperature. Then comes another detail, 39 degrees Celsius. It is a temperature, and a hot one. Finally, you are told that the temperature reading came from the mouth of a human being. That person is ill, running a temperature above 102 degrees Fahrenheit. "Everything you added dramatically changed your understanding," says Sam Adams, a research scientist at IBM, as he finishes off his brief thought exercise.

His point is the power of data in context. The data accumulates to paint a more detailed picture that becomes knowledge. That is the march toward understanding with data. The new resource of data in great volume and great variety is helpful and necessary, but the real payoff comes, as Adams puts it, from "connecting the dots" in ways that bring to light valuable insights or discoveries.

These data connections come in different genres with different characteristics, strengths, and challenges. The first is correlation—

typically, some data pattern is linked to some action or behavior in the real world. Exploiting correlation is the first wave of the big-data phenomenon, and it can be extremely powerful. Indeed, useful and profitable observations increasingly do come from "listening to the data" to find correlations. A handful of large corporations have been at this for years, using their own data. A canonical example of this kind of data discovery is the Pop-Tarts-and-beer case at Walmart from a decade ago. The giant retailer, mining the historical purchasing data from its stores, found that consumers in the path of a predicted hurricane bought strawberry Pop-Tarts at seven times the usual rate and the best-selling item of all before a hurricane was beer. Walmart's store managers don't care why that purchasing pattern occurs. They're just going to stock up on beer and strawberry Pop-Tarts when hurricane warnings come their way.

Today, this data-discovery game has opened up well beyond a select few big companies with deep pockets, vast pools of proprietary data, and corps of analysts to pore through it. Low-cost computing and software, combined with the explosion in data on the open Web and elsewhere, means small companies and start-ups can play as well. Consider start-up ZestFinance. Its cofounder and chief executive is Douglas Merrill, the former chief information officer of Google. Merrill has shoulder-length brown hair, a silver stud earring in his right ear, and on his left forearm, past the elbow, a sinuous tattoo of a peacock. ZestFinance is based in Los Angeles, but we talked over lunch in New York during a conference on big data, where Merrill spoke and I moderated a panel. Merrill's résumé testifies to his intellectual range—a PhD in psychology from Princeton and stints as a researcher at the Rand Corporation and as a senior vice president at Charles Schwab, in addition to Google. His company's creation story starts with a call from his sister-in-law, Victoria, who needed new snow tires to get to work and was out of cash. Merrill asked what she would have done if she had not been able to reach him, and he recalls that she told him that she would have taken out another "payday loan."

Merrill knew something about finance and risk assessment, but her remark sent him researching the payday lending market—loans made to people with jobs, but with poor credit ratings or none at all. At any given time, an estimated twenty-two million payday loans are outstanding, and the fees paid by payday borrowers amount to $8 billion a year—a lot of money from those in the working population who can least afford it. Merrill saw a market in need of greater efficiency and a business opportunity that would provide the social benefit of lower costs for borrowers in the subprime consumer market. The missing ingredient, Merrill concluded, was Google-style data analytics. "Underwriting, Meet Big Data" is the ZestFinance corporate motto.

A typical payday loan, Merrill explains, is for a few hundred dollars for two weeks, and rolls over ten times, or twenty-two weeks. The fees are all paid first, with the principal due at the end—a cycle that repeats every time the loan is rolled over. In a traditional payday loan, he says, a person pays $1,500 to borrow $500 over twenty-two weeks.

Using ZestFinance, Merrill says, a borrower generally pays $920 to borrow $500 for twenty-two weeks—still a lot but far less than a standard payday loan. The payments along the way pay off both interest and principal. There is no big "balloon" payment at the end. The better deal for borrowers, Merrill says, is made possible by ZestFinance's data-sifting algorithms, which reduce the risk of default by 50 percent compared with a typical payday loan. ZestFinance assumes less risk, so it can charge less and still turn a profit on its loans.

To make its risk assessments, ZestFinance's machine-learning models work on tens of thousands of data "signals" in seconds. Its data, Merrill says, comes from many sources, including the

Web, third-party information brokers, and credit services. One unconventional data point is how long a person has had his or her current cell phone number (the longer, the better). Another is how a potential borrower types his or her name into websites— all uppercase letters (least likely to repay), all lowercase (a better risk), and proper case (the best risk). Still another: about 10 percent of the borrowers approved by ZestFinance's algorithms show up on credit-bureau reports as being dead.

These walking-dead borrowers are less likely to default on their loans than the average among those approved by ZestFinance. It may be, Merrill says, that these people withstood some bad event or bad judgment and lived outside of society's computer-tracked grid for a stretch—generating no data points at all, and thus were assumed dead by conventional analysis. Having survived rough times, these people, Merrill says, may have redoubled their efforts and become better credit risks. Then he pauses and says, "You can spin a story to explain why. But you don't really know."

ZestFinance is in the business of correlations rather than explanations. These are correlations that, taken as a whole, do point to better credit risks in the subprime consumer market. And that's plenty good enough for ZestFinance, for the lenders who use its underwriting technology, and for people like Tara Richardson of Hazelwood, Missouri. To the degree that there is a representative payday borrower, it is a single working mother in her thirties. Richardson is in her thirties, was a single mother for nine years, and remarried in 2011. For the last thirteen years, she has been an elementary school teacher, mostly in public schools. She has worked steadily, other than for brief periods when she was laid off due to local budget cuts in the St. Louis suburbs. But during a lengthy and costly child-custody battle with her ex-husband, she says, more than $17,000 in legal bills piled up, and the debt eventually pushed her into personal bankruptcy.

Over the years, Richardson had borrowed from traditional payday lenders at times. But their business formula of interest fees first and a balloon repayment of loan principal at the end of two weeks seemed a bad deal to her. "If you don't have $500 today, you're not likely to have $600 in two weeks," she says, referring to the principal on a $500 loan plus $100 in interest fees. "So you have to keep going back to them. It's a vicious cycle." Yet a loan using the ZestFinance alternative, she says, involved payments of $95 every two weeks—split between interest and principal—until the loan was paid off. The total cost was hundreds of dollars less than a conventional payday loan would have been, she estimates. Richardson, her husband, and their two children were moving into a slightly larger rental property at a time when her husband's hours as an assistant manager at a fast-food restaurant had been cut. The loan, she says, was "a little extra to get us over the hump."

So correlation rules—but not always. One of the most celebrated examples of the power of correlation has been Google Flu Trends. Begun in 2008, the service monitors flu-related search terms and seeks to predict the incidence of flu, a couple of weeks ahead of official statistics. Google Flu Trends is a clever research project, a data-driven early-warning system for public health. Google search queries, going back years, were matched with government statistics based on doctors' reports to the Centers for Disease Control and Prevention in the United States. (Now the flu trend service is available in more than two dozen countries.) Google's algorithms identified searches like "cold and flu remedies" and "flu symptoms" and "influenza complications" that spiked in regions shortly before flu outbreaks were reported to health authorities. Google Flu Trends' predictions depend on tracking the frequency of search terms most strongly correlated with peaks in flu reports in the past.

In 2009, the Google service was prescient, spotting the spread of the H1N1 flu virus accurately and ahead of official reports. It was hailed as proof of the wisdom of correlation in the big-data world. But in the 2012–13 flu season, Google's algorithms stumbled. As an article in the science journal *Nature* first noted in February 2013, Google Flu Trends reported that nearly 11 percent of Americans were ill at the January peak—nearly double the 6 percent reported later by the Centers for Disease Control and Prevention. Apparently, news reports and social-media messages warning of a harsh flu season prompted a surge in flu-related searches, even if the fears of flu turned out to be exaggerated. Later in 2013, Google announced an update of its flu service code, intended to filter out the news and social-media effect. But in an article in *Science* in March 2014, four quantitative social scientists found that Google's flu-tracking service had consistently overestimated the number of flu cases in the period they studied, a period of more than two years, ending in September 2013.

Their article, "The Parable of Google Flu: Traps in Big Data Analysis," declared that Google was guilty of "big data hubris," which the authors defined as the implicit assumption that big-data sets trump traditional data collection and analysis. In a follow-up paper, the authors looked at the 2013–14 flu season, after Google had updated its algorithm in October 2013. There was some improvement, the authors found, but Google's flu-tracking service still overshot by about 30 percent. In fairness to the creators of Google Flu Trends, they intended it mainly as a "complementary signal" rather than a stand-alone forecasting tool.

Still, respected authors and academics often pointed to Google Flu Trends as proof of the triumph of the big-data approach. Tracking forty-five flu-related search terms over billions of searches, monitoring trends, and making correlations would win out. Google, it was said, could tap the "collective intelligence" of society in real time. In fact, the Google algorithms did their job. They identified correlations and quantified them. But things turned out to be more complicated and more nuanced than Google's algorithms could see. They missed *context*—some missing piece of the larger picture, which gives the data meaning; the force that "connects the dots," in Sam Adams's phrase.

Context points to a second genre of data connection that can be thought of as "association." It's a step up from correlation toward knowledge. Prime examples here are computer systems that can correctly place words in context. IBM's Watson is such a technology. A medical version of the Watson software is being fed questions from the United States Medical Licensing Examination, which every human student must pass to become a practicing physician. So the medical Watson software ought to be able to make the associative link that "39 degrees Centigrade" is a feverish temperature.

Another such technology is Google's Knowledge Graph, software that links related information in context. For example, search for "Leonardo da Vinci." On the left of your screen, you will see the standard blue links to Web articles and sites about the Italian Renaissance artist and scientist. On the right, the Knowledge Graph results: a few pictures of da Vinci, and below that a short text summary of the man, date and place of birth and death, and small images of prominent paintings including *Mona Lisa* and *The Last Supper*. Other companies, notably Apple with Siri and Microsoft with Bing, are developing their own versions of text association or knowledge software, and there are several university research projects in the field.

Our notions of "knowledge," "meaning," and "understanding" don't really apply to how this technology works. Humans understand things in good part largely because of their experience of the real world. Computers lack that advantage. Advances in artificial intelligence mean that machines can increasingly see, read, listen, and speak, in their way. And a very different way, it is. As Frederick Jelinek, a pioneer in speech recognition and natural-language processing at IBM, once explained by way of analogy: "Airplanes don't flap their wings."

To get a sense of how computers build knowledge, let's look at Carnegie Mellon University's Never-Ending Language Learning system, or NELL. Since 2010, NELL has been steadily scanning hundreds of millions of Web pages for text patterns that it uses to learn facts, more than 2.3 million so far, with an estimated accuracy of 87 percent. These facts are grouped into semantic categories—word buckets of similar meaning—including cities, companies, sports teams, actors, universities, and hundreds of others. The category facts are things like "San Francisco is a city" and "sunflower is a plant."

NELL also learns facts that are relations between members of two categories. For example, LeBron James is a basketball player (category). The Cleveland Cavaliers are a basketball team (category). By scanning text patterns, NELL can infer that LeBron James plays for the Cleveland Cavaliers, after departing the Miami Heat—even if it has never read that James plays for the Cavaliers. "Plays for" is a relation, and there are more than 900 categories and relations, and the list keeps expanding.

NELL is highly automated, and its software runs twenty-four hours a day, seven days a week. At the outset, the university researchers built a starter kit of knowledge, curated by humans, seeding each kind of category or relation with ten to fifteen examples that are true. In the category for emotions, for example: "Anger is an emotion." "Bliss is an emotion." And about a dozen more. Initially, the Carnegie Mellon team had hoped that NELL would just grind away, learning on its own, getting consistently better and better. And for the first six months, it chugged away unaided. But the researchers noticed that while NELL was very good in making many word associations, it was going badly astray on others. Now the team goes in every few weeks to correct some of NELL's most blatant mistakes, so it can learn more effectively. NELL works best with a human helping hand.

Tom Mitchell, chairman of the machine-learning department at Carnegie Mellon, offers two similar sentences as an example of what is most challenging to a knowledge system like NELL. "The girl caught the butterfly with the spots." And, "The girl caught the butterfly with the net." A human reader, he notes, inherently understands that girls hold nets, and girls are not usually spotted. So, in the first sentence, "spots" is associated with "butterfly," and in the second, "net" with "girl." "That's obvious to a person, but it's not obvious to a computer," Mitchell says. "So much of human language is background knowledge, knowledge accumulated over time." Background knowledge, then, is the missing data needed to clarify the picture, the contextual perspective.

Sometimes, the computing context-gap can be amusing and revealing. Early in its *Jeopardy!*-playing days, Watson struggled with a space-travel question: Who was the first woman astronaut? Possible answers could be Valentina Tereshkova of the Soviet Union (correct—1963) or Sally Ride (first American woman—1983). But at first, Watson consistently answered, "Wonder Woman." Just mining millions of pages of text, parsing mentions of women flying in space, by frequency and words grouped together, it undoubtedly seemed a high-probability answer to Watson, which had not yet been tweaked to separate fictional references from real-world accounts. "But we were kind of sad when Watson no longer answered Wonder Woman," recalls Jennifer Chu-Carroll, a scientist on the Watson team, as if a bit of whimsy had departed from their creation.

Wonder Woman was soon shunted aside, as Watson's knowledge base became larger, more detailed, and more refined. The computerized knowledge systems being developed at IBM, Google, other companies, and universities are starting to put together "a rich and accurate model of the world," as Daniel Kahneman summed up the virtue of human-style fast thinking. That sort of cognitive model is the engine of intuition, inference, and cause-and-effect reasoning—getting to the "why" of things, to understanding. It is a horizon of connection that is well beyond correlation. But there is a lively debate among data enthusiasts as to whether the pursuit of causes is even necessary. In their timely and authoritative book *Big Data*, Viktor Mayer-Schönberger and Kenneth Cukier forcefully state the case for correlation supremacy. "The ideal of identifying causal mechanisms," they write, "is a self-congratulatory illusion; big data overturns this."

Not everyone agrees. One of them is Richard Berner, former chief economist at Morgan Stanley. In 2013, Berner became the first director of the Office of Financial Research, a unit of the Treasury created in the aftermath of the financial crisis. Berner's office is a data-driven endeavor, an institutional recognition that leaders in the financial industry and policy makers were largely blindsided by the 2008–2009 crisis. Regulators and bankers lacked the data and analysis to see the hidden risks in the financial system. The office is staffed with economists and quants. One of the lessons of the crisis, Berner says, is that there were "serious deficiencies" in financial measurement, and new reporting requirements and data collection initiatives are under way.

Berner has to walk a tightrope. His team has to balance the need for more detailed data gathering with the need for data security, protecting the trade secrets of investment banks and other institutions, and avoid onerous reporting requirements. His Senate confirmation was held up not because of him but because of resistance from bankers and their lobbyists to anything that could add to regulatory or compliance costs. Berner is exploring tailored approaches to get a faster handle on emerging risks and sponsoring research.

One research paper proposes "privacy-preserving methods" for institutions to share information on risk exposures. The methods combine data analysis, financial economics, and computer science. Andrew Lo of MIT and his two coauthors contend that new streams of financial data—aggregated, properly encrypted, and then analyzed—could give strong clues to hidden risk bombs in the system, like the institutions that touched off the crisis in the fall of 2008, Lehman Brothers and the American International Group. Such data, the authors argue, could "have played a critical role in providing regulators and investors with advance notice of AIG's unusually concentrated position in credit-default swaps, as well as the exposure of money market funds to Lehman bonds."

This is big data as a financial microscope. The goal is to see the inner workings of markets in illuminating detail to inform understanding and guide action. So Berner is a big-data proponent, but not without qualification. He is skeptical of the uncompromising data-ists who celebrate correlation as plenty good enough without theory, without a model of how the world works. "I could argue," Berner observes, "that mentality is what led us into trouble in the financial crisis." The naïve assumption that housing prices would only go up was largely based on a blinkered view that analyzed data in more recent years, when data was plentiful and consistent. It ignored earlier financial crises, when data sets were sparse and messy.

History and its lessons—the patterns of the past—were given

short shrift. "I think the people who say that correlation is good enough should think again," Berner says. Both data and theories or models of how the economy behaves, in his view, are essential to understanding. The current debate, Berner adds, revives an old one in economics, pointing to a 1947 article, "Measurement Without Theory," by Tjalling Koopmans, a Dutch-American economist who later won a Nobel Prize. The Koopmans article was a critique of the hard-line "empiricist" approach to the study of business cycles back then.

F ew people have wielded the power of data with so dramatic effect as David Ferrucci. He led the IBM research team that created Watson the *Jeopardy!* winner. That contest ended with Ken Jennings, the all-time champion on the TV quiz show, writing on his video screen, in a gesture of genial surrender, "I, for one, welcome our new computer overlords."

The human face of Watson, to the extent that there was one, was Ferrucci, a goateed computer scientist who was always articulate and at ease in front of a camera or microphone. Yet at the end of 2012, Ferrucci joined Bridgewater Associates, a giant hedge fund, after what he describes as "a great, great career" at IBM, spanning twenty years. The weight of his Watson-related celebrity had an influence. "I was so linked to the Watson achievement," he explains, "that I felt I was almost losing my identity." Working for a hedge fund, he concedes, was never part of his career plan. But the appeal of working in a smaller environment in an entirely new field for him—applying artificial intelligence techniques to modeling the economy—won him over. Bridgewater strives to combine theory with data in a conceptual framework that the investment firm's founder, Ray Dalio, terms the "economic machine."

That approach to investment, says Ferrucci, is in sync with his current thinking in what he calls "my 30-year journey in artificial intelligence." Decades ago, the main focus of artificial intelligence research was to develop knowledge rules and relationships to make so-called expert systems. But those systems proved extremely difficult to build. So knowledge systems gave way to the data-driven path: mine vast amounts of data to make predictions, based on statistical probabilities and patterns. Data-fueled artificial intelligence, Ferrucci says, has been "incredibly powerful" for tasks like natural-language processing—a central technology, for example, behind Google's search and Watson's question-answering. "But in a purely data-driven approach, there is no real understanding," he says. "People are so enamored with the data-driven approach that they believe correlation is enough."

For a broad swath of commercial decisions, as we've seen, correlation is sufficient, as long as the outcome is a winner. But in decision-making realms like business strategy and economic policy-making, where the stakes are higher, Ferrucci says, correlation alone will never suffice. "You can be a math genius, but that's not enough," he says. "You're going to have to be able to address the question: What's really going on?"

The future of artificial intelligence, according to Ferrucci, involves a closer working partnership between data and the animating ideas of cause and effect—theories, hypotheses, mental models of the world, the "why" of things. Technical advances are making the symbiotic relationship increasingly practical. "The data informs the model," he says, "and then you have that back-and-forth cycle of improvement."

Others share Ferrucci's view. In a 2013 research paper "Why ask Why?" Andrew Gelman of Columbia and Guido Imbens of Stanford argue that the tools of measurement can be a crucial "part of model checking and hypothesis generation" in the search for causes. Peter Norvig, an artificial intelligence expert and a research director at Google, is often cited as a data supremacist. He coauthored with Google colleagues an influential essay titled, "The Unreasonable Effectiveness of Data," which made the case for the primacy of data. "Invariably," they wrote, "simple models and a lot of data trump more elaborate models based on less data. . . . So follow the data." Yet the basic point Norvig and his colleagues were making was that the social sciences do not yield to succinct mathematical theories, as in physics. Instead, progress comes from embracing "the best ally we have"—big data. "But to be clear," Norvig wrote in an explanatory blog post, "the methodology still involves models. Theory has not ended, it is expanding into new forms."

The virtue of the measurements-plus-models approach was evident well before there were computers. In his "Measurement Without Theory" essay, Koopmans cited the collaboration, starting in 1600, between the early astronomers Tycho Brahe and Johannes Kepler. They were exploring the movements of the planets, and Koopmans pointed to their work together as an ideal example of "a case where the empirical approach paved the way for the discovery of fundamental laws." When they began, the prevailing theory, firmly held by Kepler and others, was that the planets orbited the sun in circular paths. In their partnership, Tycho was the data expert, accumulating careful measurements from his state-of-the-art seventeenth-century observatory. For his part, Kepler was the flexible thinker. His "outstanding success," Koopmans wrote, "was due to a willingness to strike out for new models and hypotheses if such were needed to account for the observations obtained." The data strongly suggested that planets orbited the sun not in round circles but in more egg-shaped, elliptical flights through space.

So the hopeful prognosis for the big-data era is that the partnership of measurement and models may be renegotiated a bit, but will remain essentially intact. The computers will be crunching data, while humans do the higher-level thinking—supplying the conceptual ideas, rules, and judgment that guide the automated data analysis and prediction. The software is getting smarter, but it's not that smart. The implicit bargain, and division of labor, between computer and man remains intact.

That seemingly natural division of labor was famously articulated more than a half-century ago by J. C. R. Licklider, a Harvard-trained psychologist and seminal thinker in computing, who sponsored a wave of pioneering computer research in the 1960s as a senior official at the Pentagon's Advanced Research Projects Agency. In 1960, Licklider wrote "Man-Computer Symbiosis," a paper that would shape thinking for decades. In it, Licklider stated that the appropriate goal of computing was to "augment" human intelligence rather than substitute for it. This is technology as assistant, benign and unthreatening. It is the near-term prospect for big data, and likely to be for years. And, in truth, these technologies have to mature, drop in price, and become easier to use if big data is going to realize its potential as the equivalent of a smart assistant across a range of industries.

Yet there is another view. You hear it from scientists, researchers, and academics rather than business people. They say we're on the cusp of a historic shift. The collaboration between man and computer has so far been a partnership in which the human is the senior partner. That assumption, they say, may no longer hold. They point to artificial intelligence systems like IBM's Watson as forerunners. Murray Campbell, a slender, sharp-featured Canadian computer scientist, goes back further, to IBM's Deep Blue computer, which beat the world chess champion Garry Kasparov in 1997. Campbell was one of the leaders of the team that built

Deep Blue. Today he is a senior staff member of what used to be the mathematical sciences department in IBM's research labs and is now, in a bow to the times, called the business analytics and mathematical sciences department. "We just call it BAMs," Campbell quips.

These days, a revisionist view of Deep Blue's triumph exists among some artificial intelligence experts. A chessboard, they say, is an eight-square-by-eight-square grid—a mathematically bounded space, the favored terrain of computers. The amazing thing, they say, is not that Kasparov lost but that humans did so well for so long against chess-playing computers. Campbell has heard this retrospective carping before. In reply, he notes casually that the number of possible positions in chess is 10 to the forty-third power—written out as a number, the numeral 1 followed by forty-three zeroes, or 10,000,000,000,000,000,000,000,000,000, 000,000,000,000,000. Deep Blue mastered a formidable big-data challenge in its day.

In the 1990s, Campbell says, a major obstacle was getting chess-playing knowledge into the computer. The IBM researchers hired a chess grandmaster, Joel Benjamin. They essentially debriefed Benjamin and translated his knowledge into computer code. It was a painstaking process, Campbell recalls, because of the limited automated learning abilities of computers at the time. That computing weakness is known as the "knowledge bottleneck" in the field of artificial intelligence, Campbell explains, and it is a principal reason that progress in early so-called expert systems was handicapped. But the ability of the software in systems like Watson and Google's Knowledge Graph to build databases of knowledge with little human help, almost autonomously, their algorithms scanning vast stores of digital data at lightning speed, changes the game. "With big data and machine learning, the knowledge bottleneck is no longer the problem it once was,"

Campbell says. In fact, he predicts, "The knowledge bottleneck will reverse."

The implication, according to Campbell, is that many fields will see a rerun of chess and *Jeopardy!* The computer systems make gradual progress at first, but eventually become superhuman. The smart software will continue to construct ever-larger databases of knowledge, putting words and ideas in context and making inferences. The achievable aim, says the IBM scientist Sam Adams, is to mimic cause-and-effect reasoning—"experiential learning at scale," as he puts it. The future, Adams predicts, will be one of "machines augmented by human experts." Adams adds reassuringly that humans will still make the rules. But his comment suggests the Licklider vision being flipped, from computer as assistant to the computer assisted—or augmented, as Adams echoed Licklider—by humans.

What's wrong with that? Let the machines do more and more of the work. The march of automation, after all, is a common theme in the economic history of the industrialized world. In the early 1940s, nearly 40 percent of the American workforce was employed in factories. Today, the manufacturing share of the labor force has declined to about 8 percent, even as the nation's manufacturing output has increased sharply in value over the decades.

Yet even techno-optimists have second thoughts as they see smarter machines likely to take on cognitive tasks long reserved for humans—when what is being replaced is not sweat but synapses. In *The Second Machine Age*, Erik Brynjolfsson and Andrew McAfee of MIT make the case for a technology-led surge in productivity and growth in the future, but one that will have more sweeping and disruptive effects on society than previous waves of automation. The book, published in 2014, calls for adaptive changes in policy, education, and skills training to prevent more and more workers from being left behind. Their book also raises a central issue: As intelligent computers make more decisions, might humans lose control? To decide, after all, is to wield power.

So when do you hand the decision over to the machine-learning software? Context, once again, is crucial—namely, the setting in which a machine-made decision is being made. To explain, I think of a conversation with Claudia Perlich, the chief scientist of Dstillery, a data-science start-up in New York that specializes in ad targeting. Perlich is a former research scientist at IBM, a winner of prestigious data science contests, and a lecturer at New York University's Stern School of Business. When I ask why she is using her skills to deliver ads, Perlich replies that digital marketing is a large, real-world testing ground where practitioners in a young field can safely learn valuable lessons. The online advertising marketplace, she says, is "a wonderful place for data scientists to experiment now. What happens if my algorithm is wrong? Someone sees the wrong ad. What's the real harm? It's not a false-positive diagnosis for breast cancer."

In high-stakes decisions like diagnosing cancer, you unquestionably want a human in the loop. But systems like IBM's Watson will increasingly plumb data, accumulate knowledge, and build models of the world—on their way to sometimes becoming superhuman decision makers. How do you really control them?

At a research conference at IBM's Watson lab, Danny Hillis, an artificial intelligence expert and cofounder of Applied Minds, a technology design and research firm, took up that issue. The ever-smarter systems being made by IBM, Google, and others, Hillis says, will each need its own explanatory assistant—"a storyteller." A computer system, no matter how clever, he said, will not get very far if it just spouts answers.

The machine, Hillis explains, must be able to "tell a story about why it did what it did. The key thing that will make it work and make it acceptable is storytelling." What Hillis calls storytelling can be thought of as an audit trail that traces the data, ideas, and inferences that went into the mix of a software-generated decision. The algorithms have to explain themselves, Hillis says, to reveal "how they relate to us—how much of this decision is the machine and how much is human."

7

DATA GETS PHYSICAL

The vines are resting. It is a November day beneath a slate-gray sky in California's Central Valley. The grape-growing season here starts in the middle of March and ends in October. A few golden-brown leaves still hang on the trellised vines, but most of the leaves are underfoot and the grapes are gone, all harvested. The vines stand in neat rows that extend to the horizon, each in its assigned place, patiently waiting. The crisp fall air smells of dry, fallen leaves and rich alluvial soil. "This is autumn in the vineyard," says Nick Dokoozlian. "Yes, it's a beautiful thing."

This vineyard in the north of the Central Valley, outside the town of Lodi, is also the site of an encouraging experiment in precision agriculture, which is essentially big-data farming. Using data from on-the-ground sensors and satellite imaging, a research team was able to tailor the ideal water and fertilizer amounts to be drip-fed to vines almost on a vine-by-vine basis. In the 2013 growing season, two sections of the vineyard, side by side, served as an agricultural petri dish. One ten-acre section was managed with the precision techniques, while the vines in the other ten-acre tract received the usual standard doses of water and fertilizer. The results showed that the data-guided system delivered a decisive advantage—25 percent more grapes produced, and wine grapes of higher quality.

The test run in California involved only grapes. Yet if the data system can be commercialized, the kind of precision agriculture used in the Lodi vineyard could be adapted to other perennial crops, which grow year after year without replanting. Perennials account for a sizable portion of the farm food basket of fruits and vegetables including apples, oranges, peaches, almonds, avocados, strawberries, broccoli, and garlic. Harnessing big data to increase food production could be vital to cope with a world of increasingly limited supplies of land and water, but more mouths to feed—a global population of 9.6 billion people by 2050, a third more than today, estimates the United Nations.

Plants are not digital. But the California project shows how it has become possible to measure, monitor, and manage a complex physical system—a vineyard, in this case—in a way that was not possible even a few years ago. Lower-cost sensing, imaging, and wireless communications make it possible. Smart software and data scientists are making it happen. It amounts to a layer of intelligence being added to the physical universe, either by attaching digital sensors or by remote sensing. So business practices associated with the consumer Internet—faster learning and experimentation—are becoming easier to achieve in newly digitized environments, like a vineyard. The Central Valley research, a side-by-side comparison of two methods of crop management, is the equivalent of an Internet A/B test of an ad or online offering, except done on a patch of the earth rather than on a website.

The precision agriculture pilot was a joint effort of two companies, IBM and E. & J. Gallo Winery. And it was in good part a collaboration between two men: Hendrik Hamann, a German physicist and researcher at IBM, and Dokoozlian, a native Californian who grew up on a small family vineyard and is Gallo's chief plant scientist. Dokoozlian sees precision agriculture as the latest step in a decades-long march, transforming a craft ruled by tradition and established practice into a science relying on objective measurement. "Nothing exemplifies that more than winemaking," he says.

Dokoozlian is a big man with a sturdy frame, thinning salt-and-pepper hair, and the tan of someone who spends plenty of time outside. His grandfather emigrated from Armenia, bought some land, and farmed it, as did Dokoozlian's father, mostly growing table grapes and grapes for raisins. Dokoozlian recalls being raised on the vineyard, working before school on the farm, so that he could play on his high school football team after school. In college, he flirted with the idea of becoming a lawyer, but soon returned to his proverbial roots in viticulture. He earned a PhD in plant physiology from the University of California at Davis and taught there for fifteen years as a professor of viticulture, chemistry, and enology. Cal Davis, as it is known, is America's MIT of wine science, and one of a few academic centers of viticulture excellence worldwide, including the University of Bordeaux in France and the University of Adelaide in Australia.

Dokoozlian, the former teacher, gave me a tutorial while touring the vineyard and his lab and traversing the Central Valley in his white sport-utility vehicle. Animated and enthusiastic, he is equally at home with details and the big picture. At one point, he describes the spectrometer-level measurement of chemical compounds like beta-damascenone, which enhances a desirable dark fruit aroma in red wine. A few minutes later, he is panning back to tell the history of grape cultivation in California's Central Valley.

Until the 1960s, Dokoozlian explains, grape growing there was mainly for table grapes and raisin production, other than small amounts for sweet, fortified dessert wines like sherry and port. But the 1960s, he says, brought both a technical achievement in vineyard science and a cultural shift in American drinking habits. Two Cal Davis plant pathologists, Austin Goheen and

Harold Olmo, devised a heat-treatment technique to kill viruses in vine cuttings from Europe, improving their yields and durability for the long, arid growing season in the Central Valley. And in the 1960s, Americans developed a taste for wine, setting off a planting boom. (Farther west and north of San Francisco, the Napa Valley experienced a winemaking revival in the 1960s, having languished since Prohibition. Boutique distilleries began to proliferate in the milder climate of the Napa region.)

The breakthrough of the 1970s was modern drip-irrigation systems, developed in Israel. Drip irrigation, Dokoozlian notes, replaced the traditional "flood" method of just pouring water in the furrows between the vine rows. The new technology conserved water and made it possible to deliver nourishment—water and fertilizer—directly to each vine. Starting in the 1980s, viticulture began to focus on, and measure the results from, vineyard design. The design experiments, Dokoozlian says, have involved the spacing and placement of vines, trellising techniques, and irrigation schedules. In recent years, there have been striking advances in sensor and data collection technology used in agriculture. For example, sophisticated sensors, trailing behind a tractor or plow, probe the ground with electromagnetic waves, collecting data every second to map the depth, texture, clay content, nitrogen, and salt levels of the soil beneath.

B ut the joint project with IBM is a significant step beyond what has been done till now. It promises to break one of the enduring laws of the wine industry, which is that wine quality and quantity are inherently inimical. You can produce wine of high quality or you can produce wine in volume, but you can't really do both. Dokoozlian seeks to challenge that article of winemaking faith, and he thinks he sees the means. "We can improve both productivity and quality in a way that was unthinkable before the era of big data," he says.

On the productivity front, Dokoozlian looks at grain crops with envy. The yield of an American cornfield has quadrupled since 1920, he notes, while California grape yields have doubled over the same span. Yet grape growing does not have the same avenues of improvement as grain crops, given the marketplace constraints. Breeding new varieties for disease resistance, for example, has helped increase the production of corn, wheat, and rice. In the wine industry, however, more than 80 percent of sales are of fewer than ten varieties including Chardonnay, Cabernet Sauvignon, and Merlot. A few new wine varieties have been bred over the years but have not caught on. So breeding, Dokoozlian observes, is not a tool in the wine industry's arsenal. Similarly, he says, the use of biotechnology to genetically engineer vines to produce more grapes is off-limits, since it would almost surely be met with consumer resistance and protests against "Frankenwine." The main engine of progress that remains is agronomics, the science of crop management. And data science, Dokoozlian insists, is "going to revolutionize agronomy."

To Americans who were college students in the 1970s or 1980s, Gallo is a name that summons images of bargain-basement drinking, cheap jug wines, and Boone's Farm Apple Wine. But today, the family-owned company has sixty brands with wines in nearly every price range, yearly sales of more than $3 billion, and it is California's largest wine exporter. Dokoozlian's lab in Modesto is a single-story building in an office park. Inside, there is a large open room with a split personality. On one side, the place looks like a traditional chemistry lab with workers hunched over with beakers, vine splicings, and microscopes. The other side has rows of computers with workers mostly studying and manipulating satellite images. Nearby are whiteboards filled with formulas and calculations.

Dokoozlian's team at Gallo was doing sensing and data research long before IBM showed up. A few years ago, his researchers figured out how to use satellite imagery to calculate in detail the amount of water being consumed across a vineyard. What intrigued Dokoozlian about working with IBM was the breadth of resources it could potentially bring. IBM's supercomputers and experts have long been at the forefront of weather modeling, working with government labs. And Dokoozlian noticed that several of the company's Smarter Planet projects involved water conservation and hyper-local weather forecasting for cities— advanced technical skills that apply to agronomics as well. But it wasn't until a senior technology executive at Gallo attended a gathering in Napa Valley, hosted by IBM, that Dokoozlian knew IBM might be interested.

Soon after, in July 2012, Hendrik Hamann was on his way to meet with Dokoozlian in Modesto. IBM was looking for opportunities in agriculture, a large industry that was rapidly becoming data-rich but also a business where IBM did not have much of a presence. Hamann, a slender, boyish-looking physicist, was a scout. He was eyeing Gallo as a potential partner in an applied research project, as part of IBM's first-of-a-kind program. These projects are investments in the future, and to win approval, they must meet certain criteria. The checklist, Hamann says, includes meshing with IBM's strategy, advancing science, and being applicable to other industries. But another key requirement, he adds, and one less easy to measure, is that the project must be a strategic imperative for the corporate partner. Over a lunch in Modesto, it was readily apparent to Hamann that Dokoozlian was on a mission. "When I talked to Nick, he had a clear vision of where he wanted to go,"

Hamann recalls. "He said, the vineyard of the future—that's what I want."

Dokoozlian laid out his vision of individualized treatment for grapevines, as if the vines were people. He told the visiting IBM scientist: "It's about the right medication and dose for the patient plant. This is exactly the same as prescriptive medicine." There is far more variation across a vineyard, he explains, than you might expect, in ground temperatures, growth patterns, and soil types. The Central Valley, he notes, was a floodplain in an earlier geologic time and the waters deposited different soils unevenly across it. Each soil type has its own chemistry and capacity to absorb moisture. So overcoming soil variation, he says, is a major challenge for vineyard management.

For his part, Hamann found the project concept appealing and appropriate for IBM, and he pushed to pursue it. Still, he had his doubts at the outset. Granted, IBM has decades of experience in weather forecasting and, recently, down to neighborhood-scale predictions to warn of potential mudslides for cities like Rio de Janeiro. Using satellite data to predict local weather and crop yields had been done before. Climate Corporation, a start-up founded by former Google engineers, applied big-data weather prediction to crop insurance so impressively that in 2013 the agribusiness giant Monsanto bought the young company for nearly $1 billion.

Yet the Gallo project went beyond big-data weather analysis and prediction. It involved not only a wider range of on-the-ground sensor data but also translating the data analysis into action—the plant-by-plant treatment of vines with tailored doses of water and fertilizer. And the timetable was unforgiving, set by nature's calendar. It would be done in somewhat less than a single growing season. "Before we started, I thought it was going to be tough, given all the moving pieces and the speed," Hamann recalls. "Did we have any idea this would be successful? We had a hunch. But experiments fail all the time."

The Gallo project tapped a cornucopia of physical data gathered by sensors. The little guys of the sensor world get all the attention these days, Hamann notes. These are the digital devices that are multiplying everywhere, from fitness wristbands to roadside motion sensors, and they are getting smaller, cheaper, more sensitive, and more selective. They deserve their star turn, Hamann says. But in the vineyard research, remote sensing from high-resolution satellite imaging was crucial. Much of it comes from government-funded programs like Landsat, whose satellites take geospatial snapshots of the earth, at resolutions from a kilometer to meters. The infrared bands, when dissected by computer algorithms, present a detailed picture of a vineyard. From space, it is possible to precisely observe the health of vines, the effects of irrigation, and the spread of pests, mainly by analyzing the size and vigor of the leaf canopies of the vines, as reflected in their light waves, the equivalent of their botanical pulse. The satellite imagery captures the fitness of a vineyard on a scale and with a degree of accuracy that only an army of people, roaming the plant rows, could match, says Dokoozlian.

The satellite imagery, though essential, is only one data source. Other sources include the National Weather Service, the Department of Agriculture, giant computerized harvesting tractors, and soil sensors. The data spans weather patterns, soil types, land elevation, and the grape-producing history of individual vines (the tractors whose vibrating fiberglass rods separate the grapes from the vines measure yields by precise GPS location). This rich brew of data is fed into statistical and machine-learning software models. The models then predict the best dosage of water and fertilizer for each vine.

Hamann did his early research on the physics of tiny things. He developed optical microscopes to study single molecules at high resolution, and he worked on the nanotechnology of computer disks. His realm was not only microscale but also contained and hermetically sealed. He was doing physical modeling in a lab setting. Today, his research is in the field of sensor-based physical modeling in the wide-open world, like the California vineyard. The step into an infinitely larger physical arena, Hamann says, is achievable and practical because of the advances in measurement and data.

Actual measurements have replaced estimates and guesswork. To explain, Hamann offers a simplified example. A maths model of the behavior of a physical system that is affected by the weather includes temperature as a "boundary condition," or constraint. In the past, a scientist would take a temperature reading or two and then estimate the later temperature values. But now, with sensors and real-time data feeds, guesswork is dropped. The predictive model is constantly updated with real measurements, and the model's accuracy and usefulness improve dramatically. "It allows you to apply these physics models to the real world—to large-scale physical systems—in a way that was not possible several years ago," he says.

The Gallo vineyard in California's Central Valley is one such large-scale physical system. The pilot project has been quite promising. The elements of data success all came together: data collection and analysis yielded an insight that was translated into action, which delivered a measurable benefit. Collection, analysis, insight, and action—those are the four necessary steps in data-driven decision making. In the vineyard trial, you see all four. But how far the big-data approach will go in winemaking is still uncertain. It was a winner as a 10-acre experiment, but can it be done across 10,000 acres or 20,000 acres? Much remains to be done in bringing costs down.

The tailored, data-directed system required extra human hand-holding to run, and costs about four times the standard approach of giving all the vines the same dose of water and fertilizer. Yet demonstration projects that apply new technology and science are inevitably more expensive than conventional practice. Dokoozlian is optimistic that the virtuous cycle of downward spiraling technology costs and the economics of scale can make the new system a commercial reality before long. The success of the Gallo-IBM experiment has already proved some skeptics in the wine industry wrong. "You're standing in the middle of something that people said couldn't be done," Dokoozlian says during the tour of the vineyard. "This project is the beginning of precision viticulture." Encouraged by the pilot project, Gallo, IBM and a couple of agriculture technology equipment suppliers are developing a larger commercial prototype of the data-guided drip irrigation system.

BM's Hamann looks beyond the vineyards and winemaking. The scientist thinks that data-enhanced models of physics can translate to new business models in many industries. Companies that market physical goods from commodities to industrial equipment, according to Hamann, can use data and analysis not just to improve their products, but also to measure the value of their products to their customers. As a result, he says, companies will increasingly be in the business of selling services based on measurable outcomes. A fertilizer company, Hamann says, could use data to customize its products and their application for greater effectiveness in individual fields, borrowing some of the data toolkit from the Gallo project. The business difference, he explains, is that the fertilizer company would then be selling something of higher value—the outcome of its chemical technology rather than an undifferentiated input (bags of fertilizer).

The same should be true, Hamann insists, for all kinds of industrial equipment—selling available service and guaranteed performance without time lost for maintenance and repairs. These physical systems—whether, say, a farm field or an aircraft engine—are complex and dynamic in that they operate in changing environments like the weather. Yet the behavior of physical systems, he explains, does ultimately conform to the laws of physics: energy, momentum, and mass.

"To solve these problems, you need physics," Hamann says. "But they are very different from trying to model the behavior of human systems where the data can be so big, so messy and the feedback loops so complex." In human systems, certainty is often more elusive, the progression from cause to effect harder to see, so detecting a strong correlation becomes the attainable goal. Not so in Hamann's realm of physical systems. "Physics," he says, "looks for the cause."

Modern machines are physical systems of the first order. And no company has more experience with designing, manufacturing, and operating big machinery than General Electric. It is America's largest industrial company, a producer of jet engines, power plant turbines, rail locomotives, and medical imaging equipment. GE makes the heavy-duty machinery that transports people, heats homes and powers factories, and lets doctors diagnose life-threatening diseases. The company, founded by Thomas Edison in 1892, resides in a different world from the consumer Internet. But it is at the forefront of bringing big machines into the age of big data.

The concept has been around for years—digitizing machines with sensors, enabling them to communicate, and tapping the resulting vast flows for new discoveries and profit-making possibilities. The idea is part of a larger vision of putting sensors—down to "smart dust"—on all kinds of objects around the globe, gathering information, and communicating with powerful computer networks. It is popularly known as the Internet of Things. The ultimate goal, according to Larry Smarr, founding director of the California Institute for Telecommunications and Information Technology, is a "sensor-aware planetary computer."

GE's more modest formulation is what it calls the "industrial Internet." For the company, the industrial Internet is a marketing term attached to a major strategic initiative, backed by sizable investment. In May 2009, just before the Great Recession ended, the economy was still weak, but Jeffrey Immelt, GE's chief executive, decided that it was time to seriously look for opportunities in the future. He met with his executive team to identify the next wave of technology that would drive industrial productivity and that GE could exploit. The answer, Immelt recalls, boiled down to big data and smarter machines. To pursue the opportunity, Immelt decided to set up a unit far from GE's headquarters in Fairfield, Connecticut, in the East Bay of San Francisco. "To drive change, you need translators," people from outside the corporate culture, Immelt explains.

GE tapped an outsider to lead the effort, William Ruh, who had been an executive at Cisco, the big computer networking company. Ruh, a burly, affable man, is a Silicon Valley veteran. His degrees are in computer science, and he has held technology management jobs at big companies and start-ups. By early 2014, Ruh had hired a staff of 800, most of them working at GE's global software and data analytics center in San Ramon, California. The plan calls for further hiring and a total investment of $1 billion in the software and data group by 2015. Across the company, GE has some 10,000 software engineers. But they are mainly specialists in the code that animates heavy industrial machinery.

In San Ramon, the engineers are developing software for data analysis and tools that can be used across the company's industrial divisions and by customers. Inside the multistory office building of steel, stone, and tinted glass, the workforce includes designers as well as engineers. To really have an impact, data tools have to reach an industrial audience that includes machine operators, field engineers, and aviation fleet managers, not just scientists. David Cronin left a San Francisco design firm to join GE. In San Francisco, he observes, so much of the energy in software design is dedicated to enhancing the usability of consumer applications on smartphones and social networks. The appeal of GE, he explains, is using design in service of health care, energy conservation, and efficient transportation. "The social impact is a big part of it," Cronin says. "It's not just design by guys in black T-shirts."

The young scientists GE has recruited express similar sentiments. When Sharoda Paul finished a postdoctoral fellowship in 2011 at the Palo Alto Research Center, she considered a job at a big Silicon Valley company, in her case, Google. But she chose GE instead. Since she joined, Paul has donned a hard hat and safety boots to study power plants. She has ridden on a rail locomotive and toured hospital wards. "Here, you get to work on things that touch people in so many ways," she says. "That was a big draw."

Yet the underlying technologies that power Google and Facebook are also vital ingredients in the industrial Internet—artificial intelligence techniques, like machine learning. And machine operators, like consumers, typically prefer to have their software delivered to mobile devices—smartphones and tablets. GE managers often describe their work with analogies to the consumer Internet. One software service, introduced in 2011, allows sensor-equipped jet engines to send updates on their status to maintenance engineers and fleet managers. "It's constant communication on where they are, where they're going, and how they feel—it's Facebook for engines," Ruh says.

A GE software application rolled out in 2013, called Predix, makes use of "contextual computing." In an industrial setting, contextual computing means gathering and analyzing data to give a machine's human minders a view of its health, its history, and its surroundings—the context in which the machine is operating. Predix is intended as an intelligent assistant that sifts through data on a particular machine to, say, alert a plant engineer that a gas turbine needs preventive maintenance—a heads-up prediction that the turbine is heading for a breakdown. It is the industrial machinery equivalent to Google Now, a predictive search service for mobile devices, including Google glasses, that presents driving directions, recommendations for nearby restaurants, sports scores for teams you follow, based on your location, your interests, and what you've done in the past—the context of your life.

GE wants to push contextual computing in the machine world to another dimension. Most of the current focus has been on gathering data from machines to learn about them—to reduce lost operating time by applying data-driven preventive maintenance. With machines, as with vineyards, the comparison typically offered is to preventive and personalized medicine. You can prevent machine failures with tailored treatment, informed by data.

But the next step is technology to enable the machines themselves to learn about their environment and adapt to it. Take the case of a jet engine on an airliner whose route takes it to different climates, like flying from Arizona to Canada. The temperatures, humidity, and air density, for example, are usually very different in those locations. Ideally, the engine should be able to adapt its operations—air intake, rotor speed, and tilt—to its environment for maximum efficiency, saving fuel and reducing wear on the engine. "Your data models are constantly updating and adapting to the environment," Anil Varma explains. "When the machines can learn from their changing context, then you have something that really gets smarter over time."

Varma, an industrial data scientist, has spent his career trying to make big machines smarter. He earned his PhD in mechanical engineering from the University of California at Berkeley, and his doctoral thesis was on the use of artificial intelligence techniques in mechanical design and diagnostics. In 1997, Varma went straight from Berkeley to GE's main research lab in upstate New York, as a research scientist. He is a coinventor on nine patents, all of them involving the analysis of data to monitor the health of machines. For seven years, Varma led GE's machine-learning research from the main lab in upstate New York. When I first met Varma, he was the chief data scientist at GE's San Ramon center. (In 2014, he left to join Schlumberger, the large oil-field technology company, and "an exciting place for data and analytics right now," Varma says.)

Varma enjoys ministering to machines, and he marvels at them. He talks of standing next to a modern locomotive and gazing down the track at a freight train—"that huge thing that goes on and on." And he believes the work is not just satisfying but important to economic progress. "The next big step up in productivity," he insists, "is going to be getting more productivity out of these complex systems of machinery."

Varma is no economist, but he is making the economic case for big data, when it is linked to the physical world. It is accepted as a self-evident truth in business circles and championed by some academics. Still, the case for optimism remains unproven. Since the big-data wave is just getting under way, its impact has not yet shown up in official statistics. Skeptics doubt that it will have a significant effect. GE, not surprisingly, is firmly in the optimists' camp. In 2013, a report coauthored by its chief economist Marco Annunziata and Peter Evans, then its director of global strategy and analytics (Evans left GE later that year) concludes that the combination of intelligent machines, big data, and changed work practices should bring "enormous economic benefits" over the long term. The effect, they say, will be to boost the global economy by as much as $10 trillion to $15 trillion over the next twenty years. Even the low end of that range would mean adding more than another present-day China to the world's economy.

Faster productivity growth, according to Annunziata and Evans, will deliver the gains. For example, they estimate that 300 million labor-hours a year are spent to service the world's power plant turbines, aircraft engines, rail locomotives, and medical-image scanners, at a cost of $20 billion annually. Most of the maintenance today is done data-blind, using rule-of-thumb standard schedules and in reaction to breakdowns. Huge savings, the authors declare, will result from individualized, predictive servicing of industrial equipment, made possible by sensors and data analysis. The result, they write, will be to move close to the ideal of "zero unplanned downtime."

But data doubters are not persuaded. And the most prominent pessimist is Robert Gordon, a leading economic historian and a professor at Northwestern University, who made his case in a research paper published in August 2012, "Is U.S. Economic Growth Over? Faltering Innovation Confronts the Six Headwinds." In his paper, Gordon asserts that the gains from computing and the Internet have petered out in the past decade. Linking modern communications to computing, he observes, brought the Internet uptick in productivity from 1996 to 2004, a relatively brief historical period. Since the early 2000s, Gordon sees technological innovation mainly in consumer electronics. Those inventions, he writes, are "smaller, smarter and more capable, but do not fundamentally change labor productivity or the standard of living" in the way that indoor plumbing, electric lighting, and the automobile did. Gordon's paper brought an outcry from Silicon Valley and technology optimists in academia. In December 2012, Gordon responded to his detractors in the *Wall Street Journal*. In the article, Gordon observes that he has been accused of a failure of imagination. "But," he writes, "I am not forecasting an end to innovation, just a decline in the usefulness of future inventions in comparison with the great inventions of the past."

Technology optimists say Gordon is missing the larger—and deeper—picture of innovation today. He seems to see smartphones merely as more compact electronic devices instead of as mobile delivery platforms for data-fueled artificial intelligence. In his paper, the term "artificial intelligence" does not appear. Smarter machines? He mentions "robots" twice. In 1961, he notes, General Motors introduced the first industrial robot. Later, Gordon writes that by the past decade, while the role of robots continues to expand in manufacturing, "the era of computers replacing human labor was largely over." Something like GE's bet on the industrial Internet—smart machines and big data—is not on Gordon's radar, for example. That is a crucial blind spot, in the view of the techno-optimists. "The reason I think Bob Gordon is wrong is precisely because of the kind of thing GE is doing," says Andrew McAfee, a research scientist at MIT's Center for Digital Business.

Yet it is not only the techno-optimists who question the Gordon stance. Former skeptics see evidence of long-simmering technologies coming to a boil in ways that could well lift growth. One of them is Paul Krugman, a columnist for the *New York Times*. A Nobel Prize–winning economist, Krugman has long been an articulate deflator of the breathless case that modern computing is revolutionizing the economy. By December of 2013, however, Krugman had become more impressed by advances in computing and he wrote an article, published on the *Times*'s website, explaining why he thinks Gordon is "probably wrong." A decade ago, Krugman writes, "the field of artificial intelligence had marched from failure to failure. But something has happened—things that were widely regarded as jokes not long ago, like speech recognition, machine translation, self-driving cars, and so on, have suddenly become more or less working reality."

Data and software, Krugman observes, have forged the path to working artificial intelligence. "They're using big data and correlations and so on," he writes, "to implement algorithms—mindless algorithms, you might say. But if they can take people's place, does it matter?" Krugman's tentative conversion is noteworthy because it comes from someone of his stature who has a deep understanding of the economy. It takes a lot to make a difference across the $17 trillion American economy, and Krugman is acknowledging that the current wave of technology—big data and smart machines—may well be a significant force. A big data–powered lift in productivity would not solve the "headwinds" problems Gordon identifies, including an aging population, income inequality, and a struggling public education system. But without higher productivity and growth, those challenges only loom larger.

G E opens a window onto the industrial economy, much as IBM does with information technology. GE's industrial Internet strategy is based on the well-informed assumption that seemingly small steps in efficiency can have a big impact—in savings and profits for companies, and in productivity for the economy. GE's machines labor in major sectors of the economy, transportation, energy and health care, where a percentage point or two of improvement translates into billions upon billions of dollars.

Its early industrial Internet projects have been aimed at using sensors, data, and computing to achieve small gains that add up— gas and wind turbines that operate a few percent more efficiently, hospital systems that can handle more patients in a year, airlines that can shave expenses by managing fuel and flights more efficiently.

For the commercial aircraft industry, a 1 percent gain in fuel efficiency would save about $3 billion a year. In its markets, GE is typically the No. 1 or No. 2 supplier; its machines are everywhere, powering the industrial economy. "Small improvements in the efficiency of our installed base of our equipment can deliver massive improvements in profitability for our customers," Immelt says. And, of course, GE plans to profit as well. The company's industrial Internet offerings add to its services business, which generates more than $40 billion in yearly revenue. Maintaining and upgrading the machines with software and services are more profitable business than selling the machines themselves. At the end of 2013, customers had placed orders totaling $800 million for industrial Internet services.

The same magnifying principle of small gains–broad reach applies to the economy as well. A 1 percent improvement in productivity growth, to nearly the rate enjoyed from 1996 to 2004, would over the next twenty years raise the average income per person 25 percent higher than if the current trend continues. That would be no mean feat, to be sure. Yet it is the level of productivity gain that the industrial Internet technologies of big data and smart machines can potentially generate, according to GE's economic experts, Annunziata and Evans. And merely because their analysis is self-interested doesn't necessarily mean it is wrong.

The GE foray into big data is instructive, for behind its strategy is a perspective on the development of data technology. First, its industrial Internet effort is less a new business than one layered on its existing business. From the outset, the corporate motivation, Immelt says, has been "to elevate and get more out of our huge installed base" of industrial machines. Second, the GE take on data-smartened machines is that they represent an immense opportunity but one that is within the mainstream marketplace rather than being a break with it.

The GE vision is essentially of the same machines doing their tasks more efficiently, thanks to a measure of automated intelligence. And GE's bet is that there are big gains in productivity to be had without driverless cars filling the roads and robots delivering packages to your door. It is true that incumbents, like GE, reflexively favor an evolutionary path. Yet it is also true that the world of big data will come sooner, across more of the economy, if GE is right.

8

THE YIN AND YANG OF
BEHAVIOR AND DATA

Y oky Matsuoka was known as a robot wizard a few years ago, so much a star that she could pick and choose what she wanted to do. At Google, she was one of the founding members of the company's X Lab, the secretive unit that became the incubator for Google's driverless cars and Internet-connected glasses. Back then, in 2010, the multitasking Matsuoka also headed a laboratory at the University of Washington dedicated to neurobotics, a term she coined to describe the marriage of neuroscience and robotics. It aspired to put a chip in the human brain that would control a lifelike prosthetic arm—a scientific breakthrough in its own right but also one that could change the lives of war-veteran amputees and others with physical and neurological handicaps. Her stature in the field grew out of a stellar academic career, beginning as a graduate student at MIT, and later as a researcher at Carnegie Mellon, Harvard, and elsewhere. In 2007, she was awarded a MacArthur "genius" fellowship.

Yet in 2010, Matsuoka joined a start-up company. Its product? Thermostats. A dumbfounding move at first glance, but Nest Labs was not just any start-up, nor did it plan to make humble household wall fixtures. Nest was cofounded by Tony Fadell, a former Apple executive who designed the iPod, and then headed the iPod and iPhone division until he left in 2009. The other cofounder was Matt Rogers, a younger Apple alumnus. They recruited an impressive team of Silicon Valley talent in hardware, software, design, and data analysis. They won the backing of blue-chip venture capital firms including Kleiner Perkins Caufield & Byers and the investment arm of Google, as well as Generation Investment Management, cofounded by Al Gore and dedicated to environmentally responsible investments.

The founders' pitch was that Nest had a historic opportunity to transform the conventional thermostat from a dumb switch into a clever digital assistant that would save home owners money, reduce energy consumption, and curb pollution. A few industrial companies sold programmable thermostats, but they proved to be so hard to program that few people did. The Silicon Valley start-up would make a digital device that didn't ask users to program it. Nest was producing "the world's first learning thermostat—a thermostat for the iPhone generation," as Fadell told me in the fall of 2011, when Nest was about to introduce its first product. In Fadell's telling, Nest was a new take on Silicon Valley's favorite story line: change the world and make a bundle.

About half of household energy consumption in the United States is heating and cooling, and most people set their thermostats and forget them, until they really notice the cold in winter or heat in summer, and then crank up the heat or air-conditioning, respectively. There is no precise measurement or science to it. Yet just 1 degree warmer in summer or cooler in winter delivers a 5 percent change in energy use. Early trials with Nest's smart thermostats found that home owners could reduce the energy needed to heat and cool their houses by 20 percent without feeling less comfortable. If the company's technology took off, it would mean burning a lot less carbon-spewing fossil fuel. Nest appealed to a group of technologists who were ready to look for challenges beyond consumer markets.

T he Nest thermostat is a stylish piece of hardware, a circle of brushed stainless steel, reflective polymer, and a crystal-sharp color display. Tap on the display and the menu surfaces. Its look, feel, and even the pristine white-box packaging echo the iPod and iPhone. It is a product designed by artists, not put together by mechanical engineers. But, most of all, the Nest thermostat is a smart data machine. What makes it a "learning" thermostat is the machine-learning algorithms that interrogate a wide variety of data streams from its sensors, weather reports, and information the user offers—combined with intelligence gleaned from the company's historical database of accumulated information on how Nest devices are used by its customers.

Nest's invention straddles the machine and human worlds. It is a complex if compact machine, but one that observes and learns from human behavior. A Nest thermostat is not an android, marching around the house; it is affixed to the wall. Still, the technology that animates its beating digital heart—machine learning applied to data—is also an essential technology in the robotics work Yoky Matsuoka did for years. At Nest, she is the vice president of technology, in charge of developing the artificial intelligence software that is its learning engine.

M atsuoka's life and her career path have consistently taken unpredictable turns. In her early forties, she is the mother of four young children. Her dark brown hair falls past her shoulders, and she is tall, with the easy athletic stride of the tennis star she once was. She grew up in Tokyo, an only child, and recalls seeing John McEnroe play tennis there at an exhibition match when she was a little girl. She was struck not just by his play, but also by his fierce individualism and his willingness to be different. It left an impression on the young girl in Japan, where cohesiveness, even conformity, still tends to be valued over individual expression.

Matsuoka began playing tennis seriously. When she was sixteen, her parents sent her to America to attend a Nick Bollettieri tennis camp in Florida. She went alone, lived with an American family, and struggled with the language at first. English is taught in Japanese schools, but the emphasis is on vocabulary and grammar, not spoken English. "For the first three months, I couldn't speak," Matsuoka recalls. She honed her English on popular American television shows, like *The Cosby Show* and *Family Ties*, watching and absorbing. Her tennis improved and she made the Wimbledon qualifying rounds.

But Matsuoka's dream of being a professional tennis player was dashed by a series of injuries including three broken ankles, a severed Achilles tendon, a torn patellar tendon, and a back strain. Still, it soon became clear in school that she had other talents; she was acing her science and maths courses and tests. For a while, that academic success made her uncomfortable—the cool, popular, tennis star didn't want to appear to be a geek. She would pretend she wasn't studying, and go out of her way to avoid being seen lugging books around the school hallways. Then, a couple of days before a test, she would find a hiding place in the library, hole up, and study. "I had to live a double life," she told an interviewer on the PBS science program *Nova* in 2008. "I never tried to stop learning math and science. I just secretly did it."

Matsuoka's self-consciousness eventually passed, and she majored in computer science at the University of California at Berkeley, and in her graduate studies at MIT. Her mind would give her a career and affluence in a way her body, playing tennis, never could. But her tennis injuries—her body breaking down—served as inspiration to think about how the human body works and how a computerized mechanism might supplement or replace a human limb or a person altogether. Tennis, oddly, led to robotics for Matsuoka.

With household heating and cooling, Nest represents a move into new terrain for the collaboration between humans and machines—what Matsuoka describes as "the yin and the yang between understanding human learning and machine learning, that combination, that intersection, is exactly where I live." The Nest thermostat has two infrared sensors—"two eyeballs," as one engineer put it—that see, in their way, at two distances. The close-up sensor detects how people are adjusting the thermostat, while the farther-out sensor picks up movement in the room.

By now, hundreds of thousands of Nest thermostats have collected enough data and Nest's algorithms have done enough analysis, based on their patterns of activity and energy use, to determine that households can be grouped into four kinds: families with young children; families with older children; empty nesters; and roommates. Within a week after it is installed, the Nest thermostat has observed enough to know what group a household fits into. But a general pattern of behavior isn't an ironclad rule. So in a household with two roommates, and one uncharacteristically stays home on a workday, the Nest distance sensor is able to notice and adjust. It reads the human presence as if a message, "I'm home, don't flip to the energy-saving setting."

A learning thermostat, Matsuoka soon found out, should be smart but should not be perceived as arrogant. And the Nest device got a personality makeover, even before it was introduced in the fall of 2011, based on the reactions of prototype testers. Initially, Matsuoka wrote the software so that after the person dialed in target temperature ranges, the learning algorithms took over. If Nest's smart software discerned more energy-efficient settings, up or down a couple of degrees, the thermostat did so automatically. Many people hated that, she recalls; they didn't want a machine to be in control. So she took a page from behavioral economics, nudging people to make better choices with encouragement. The Nest thermostat offers a visual reward—a green-leaf icon on its display, when the user chooses an energy-saving setting. The green-leaf rewards are individualized, based on local weather conditions and a household's history of energy use. "If you keep chasing the leaf, you have a very nice schedule," Matsuoka says.

The green-leaf solution tells us something about the uneasy alliance between people and computers. The people who tried out the prototype thermostats—mechanical autocrats that set temperatures on their own—felt the machines had taken over. They rebelled at the sense that they had lost *control*. The natural order, they sensed, had been upset; humans should be the rulers, computers the assistants. But in this emerging era of big data and smart machines, the issue is going to be less about control than about *trust*. That is, what are the conditions under which you feel comfortable letting data-fueled algorithms take over? When do you let them make the decision?

Today, for example, nearly all the people using Nest thermostats do let the machine automatically make temperature changes to save energy. The company has designed a get-to-know-you period, where people put in ranges of temperature settings they find comfortable and the learning thermostat studies a household's energy-use patterns. But soon, Matsuoka's algorithms are in control more than they are not. People check in on temperatures and energy savings nearly twice a day on average, from smartphones and Web apps. The human users are interested partners and can override the machine, but most of the time they let the Nest algorithms take over.

The issue of when to trust the machine—a mechanical one or a virtual one, a software algorithm—is going to play out repeatedly in the future. Appeals to efficiency alone will not carry the day. Advocates for self-driving cars marshal safety statistics and logical-sounding arguments to push their case—about accident rates and the human foibles of drowsiness, distractedness, and drunkenness. Those arguments help, but they do not speak to the issues of trust and comfort with the machines. People are not aggregates; we all experience the world as individuals. So declaring that something will be good for the population, on average, isn't entirely persuasive. What will be needed is the storytelling that Danny Hillis, the artificial intelligence expert, describes as the machines explaining themselves, giving a simplified account of how they work. What is also needed is time—a threshold of accumulated experience of living with the decision-making machines, in the house or on the road, to reach a level of comfort.

That sort of human-machine accommodation is far easier to reach with a learning thermostat than with a self-driving car, of course. Part of Nest's allure for technologists is that it provides an avenue for putting an artificial-intelligence device into mainstream use today without the need for new laws and regulations. Mark Malhotra, a young Stanford-educated engineer, worked at Volkswagen's Silicon Valley research lab and then at Matsuoka's neurobotics lab at the University of Washington before coming to the Nest in 2012. Doing something that could have an immediate effect influenced his decision. "A lot of robotic work is further out, and it will really affect people in ten years or so," Malhotra says. "Here, the things that I work on have a real impact on people and their energy bills, and it is important to the environment and the world."

Malhotra's comment is commonplace at Nest's main office, a low-slung two-story building in Palo Alto, California, with an overflowing parking lot and willow trees out back. The sense of its larger mission permeates the place. Yet another common thread that comes through, in talking to its engineers and executives, is how much Nest is a data business. Malhotra personifies the point. A trained mechanical engineer, Malhotra is a data science practitioner at Nest. His previous robotics research, he says, pushed him in that direction. At the University of Washington, when working on a robotic hand, for instance, the mechanical electronics were important. "But controlling it, getting it to do what we take for granted, is the really hard part," he explains. That involved writing software to handle the data signals from sensors on the prosthetic hand and writing machine-learning software to train it. At Nest, he calls himself an algorithms engineer, and he is adept with the software toolkit of data science. His job is to write the software that makes sense of the sensor and other data to school its learning thermostats. He describes it as "thermal modeling in the home."

T ony Fadell, the chief executive of Nest, is at heart a hardware designer who loves elegant objects of utility. The Nest thermostat is such an object, as is the smart smoke alarm and carbon monoxide detector that the company introduced in late 2013. It communicates in spoken words, offering information and suggestions, instead of emitting earsplitting beeps. Yet Fadell knows that data handling is what makes his products truly distinctive. It is the data off the thermostats' sensors that fuel Nest's learning algorithms, both the ones running remotely in a data center "cloud" and those running on the device on the wall.

Rationing and optimizing the data streams are engineering feats. The Nest thermostat communicates using a household's wireless Wi-Fi network, so it sends its messages in a calibrated stream of data rather than a fire hose. "You don't want to hog the user's bandwidth," Fadell explains. "The last thing you want is the equivalent of a video stream from a thermostat. The response time would be lousy, and everyone would be mad at you." Nest has found that some data is useless. In some households, couples fight over the thermostat, changing the settings, usually from smartphone apps, every half hour or so. "We throw out the data that looks like internecine family war," he says. "We're not marriage counselors."

Still, most of the data goes to inform Nest's algorithms. It's a continuous loop, Fadell says, that goes into steadily improving the product and service to make the thermostat more efficient and more appealing to users. He contrasts that with his days managing Apple's iPod business. The digital music player, he notes, was a stand-alone electronic device, not continuously connected to the Internet. So the feedback mechanism for information, he recalls, consisted of market surveys, customer support calls, and e-mail complaints.

The smart-machine loop, Fadell explains, is dramatically more powerful, informed by immediate data on the product's performance and users' preferences. "It's just like the scientific method, done in real time," he says. Fadell talks of conducting A/B experiments, as Google and Facebook do, to test what customers like and don't like. The only difference is that Nest is doing so with a product that bridges the physical and Internet realms.

Not far from the Nest building, in his office at the venture capital firm Kleiner Perkins, Randy Komisar picked up on that same theme. Komisar, like Fadell, has a shaved head, and he has the physique of someone who spends serious time cycling on less traveled roads in Silicon Valley. The two men are cycling buddies, and Komisar is the Kleiner partner who led the firm's investment in Nest.

I n Nest, Komisar sees striking evidence that the business practices made possible by data, which originated on the Web, can be applied more broadly—to transform industries, in some cases. "It started with the pure Internet companies getting better metrics and instantaneous feedback to learn what's going on," he says. "But now that's spreading everywhere. Big data is the next stage." Using data to measure, test, and understand, he predicts, will play an ever-larger role in the development of products, companies, and business leaders. "Value," Komisar says, "will increasingly come from being great at reading the tea leaves in the data."

In January 2014, Google bought Nest, a start-up with 300 employees, for $3.2 billion in cash—a rich payday, though Nest's founding team was already affluent, by any normal standard. For Google, the purchase is hardly a bargain, but it's almost pocket change for the Internet giant. The deal was about money only to the degree that all transactions are; the principal motivation lay elsewhere. From the outset, Fadell called Nest's learning thermostat a starting point toward a vision of "the conscious home," as he puts it. Nest was growing apace on its own, but Google's deep pockets and global reach could accelerate things.

Google and Nest have a lot in common. They use the same underlying data-analyzing and machine-learning technologies, and they have employed some of the same people, like Yoky Matsuoka. And their worlds of the physical, digital, and data are steadily merging. Both companies are also engaged, broadly, in the same endeavor. For Nest, it is in the home to save energy; and for Google, it is on the Web to improve search and sell ads: they both observe human behavior.

In another industry, Michael Haydock uses data to observe human behavior in a very different way than Nest. His instruments of measurement do not include a direct digital connection to a physical device in households, as Nest enjoys with its learning thermostat. Haydock's data observations even lack the clarity of Google's since when a person types in a search term, it is a straightforward statement of interest. Google is a "database of intentions," as the author and media entrepreneur John Battelle put it. Haydock operates at a further remove, as he culls data to observe, and predict, the behavior of people that retailers are trying to sell to. From data, he tries to coax information on customers and prospective customers, on their lifestyles, desires, and product preferences. His applied research is then used to help tailor marketing campaigns.

As a popular taste and trend spotter, Haydock casts an unlikely figure.

A distinguished engineer at IBM, he is in his early sixties, a big man, a bit thick around the waist, but nimble and agile. He lifts weights and jokingly says he aspires to be "the world's strongest quant." He picked me up one drizzly Saturday morning at a motel in suburban Minneapolis, wearing a fleece jacket, shorts, flip-flops, and a broad smile.

Haydock lives in Chanhassen, Minnesota, a suburb about twenty miles southwest of Minneapolis. His home is a trim, modern house, with a lot of exposed wood and comfortable leather furniture, near a small lake. Mounted on a downstairs wall is a swordfish he caught (he keeps a fishing boat in Florida and has a condominium in Silicon Valley). On the wall in his study is a patent that bears his name as an inventor: "System and method for increasing the effectiveness of customer contact strategies," reads the summary. But Minnesota is mainly Haydock's base camp these days. He flies out most Sunday afternoons to work on projects with clients, returning Thursday night or Friday.

After a day spent with Haydock, at his home and over a walleye dinner at a downtown restaurant, listening to him describe his craft, I couldn't help but think of the contrast with Cayce Pollard, the protagonist in William Gibson's 2003 novel *Pattern Recognition*. She is a young marketing savant, a cool hunter. Her typical attire is a shrunken cotton T-shirt, worn with black jeans, boots, and a bomber jacket. She possesses, Gibson writes, "an unusual intuitive sensitivity for branding."

Haydock may not be hip or intuitive like Cayce Pollard. But he brings something else—data science. His PhD is in operations research, which applies maths and statistics to complex decisions. The field got its start in World War II. Operations research practitioners used quantitative analysis to reduce the number of ships lost to German U-boat attacks and to improve the accuracy of B-29 bomber raids on Japan. Later, corporations embraced operations research in industries like petrochemicals and airlines, and in managing suppliers to cut costs and improve quality. Walmart, for example, is renowned for overhauling its supply chain with statistical science. It is only in recent years, Haydock says, that similar techniques have been used to "dissect elements of human behavior" to make marketing less an art and more a science.

Haydock speaks of a new "genomics of business" in the future that will produce a previously unimagined "level of detail in looking at people and companies." Today, it seems an aspirational analogy, but that is the direction things are heading. Haydock's work is evidence of the trend. In a project for a retailer, the data might well range from hourly weather data and local gas prices to tens of millions of social-media posts, as well as months' worth of in-store and online purchase data. These are grouped into hundreds of customer categories, which can be then diced into further categories—in the thousands—based on "signals," like a change in a relationship status update on a person's Facebook page. With more data, more measurement, "the groups just get smaller and smaller," Haydock says, moving closer and closer to personalized marketing messages. If you add in the records from credit card and store card purchases—data trails that really do identify individuals personally—the privacy implications of this laser-beam marketing loom large. It is a business and public policy issue that Haydock brings up on his own, unbidden. "I have the ability to be as creepy as you can possibly imagine," he observes. (Privacy and surveillance in the age of big data is the subject of a later chapter.)

H aydock's tiny, data-derived customer groups are microscopic—a qualitative change in commercial observation—compared with the broad customer segments defined by the techniques of traditional marketing. One of the most influential was a methodology for identifying different customer markets that came out of SRI International called VALS, for "Values, Attitudes, and Lifestyles." The system was developed in the late 1970s by a team at the research institute, led by the sociologist Arnold Mitchell. It drew heavily on the earlier research of the sociologist David Riesman, author of *The Lonely Crowd*, a study of modern conformity, and on the work of the psychologist Abraham Maslow, who viewed behavior as the pursuit of a "hierarchy of needs." VALS classified people according to nine personality types, including achievers, belongers, sustainers, and societally conscious. Consumer behavior, then, was explained by a mere nine categories.

The customer projects Haydock works on are short in length,

usually four to eight months, and sharply aimed at a specific challenge the company is struggling with. When he departs, Haydock wants to leave insights and a path to future progress, some working technology, and a group of data believers inside the company. "It's a really focused, skunkworks approach," he says. "We don't want a research project that never solves anything." Haydock essentially leads a big-data SWAT team on each project, with a core of three people and a handful of pickup members with industry expertise. The permanent trio, besides Haydock, includes Paul Riedl, a crack programmer Haydock has known for twenty years; and Kevin Keene, a young data analyst who is a social-media expert.

In selecting team members, Haydock says he looks for technical skills for starters, but more important are two characteristics, curiosity and persistence. His own career has shown plenty of both. He spent years at Control Data Corporation and IBM, mostly in sales and management positions, and yet taught himself to program. Later, he was briefly the president of Cray, a supercomputer company, and then an independent consultant. Haydock got his PhD when he was fifty-seven years old, after studying for nights and weekends for more than four years. In 2009, Virginia Rometty, then head of IBM's services business, called him and suggested he return to IBM and join its fast-growing data analytics group. Haydock said he would, with one condition. "I'd be a practitioner, not an executive," he recalls.

Haydock is called in to explore the future—new market opportunities—rather than using data for cutting costs. Retailers have long experience with data collection for improving efficiency, starting four decades ago with the introduction of the bar code, the rectangular thicket of slender bars and spaces on products. It arrived when the computer revolution was beginning in earnest. Bar codes were the information-rich sensors of their day. Their data sped through scanners, computerized cash registers, and mainframes to automate broad swaths of retailing for decades. In that round, data technology helped reduce labor costs, change relations between manufacturers and retailers, and hasten the rise of efficient mass-merchandisers like Walmart. Yet most of that data was captive, from sources inside a company's internal network, from its stores to its suppliers. It was the pre-Internet era of data mining. Today, the potential data sources are obviously far more abundant, but finding intelligence in the digital babble is the quandary.

Enter Haydock and his data team. When I met him in Minnesota in the fall of 2013, Haydock had recently finished a project in New York and had begun making weekly shuttle-trips to Seattle. Haydock's data projects involve commercial and competitive secrets, and the two companies would say no more than to acknowledge that IBM was working for them. The assignment in New York was for Macy's, to help the department store retailer decipher and attract millennials, the demographic bulge of young consumers, from sixteen to thirty years old. In Seattle, the project was for Starbucks, to better understand its customers as it tries to increase food sales in its coffee shops. Respecting the code of client confidentiality, Haydock declined to discuss the details, findings, and recommendations of those projects. But he did describe how he practices data science.

H aydock views much of traditional marketing and sales as an information failure. A person enters your store or visits your website, he says, to learn about products or to buy. Ideally, that person's experience should be tailored to his or her interests or needs. But usually, the retailer resorts to "pitching products," he notes, because of a lack of information to offer the more personalized treatment. The very wealthy have personal shoppers, and the dream of big data is to digitally democratize that sort of bespoke experience. If that ambition can be achieved someday, it will be because of a core tenet of data-ism: that at some point the *quantity* of data that can be gathered and analyzed makes it possible to deliver a *qualitative* change—in the experience of marketing and shopping, in this case.

In the social-media part of the Macy's project, for example, 40 million messages on Twitter, Facebook, and blogs over three months were gathered—by or about eighteen- to thirty-year-olds. The 12 million authors of those posts were winnowed to 2.3 million as having the most relevant text. The text phrases have to be grouped into hundreds of different concepts discussed and conversational types. A key step is building the model that associates words, or clusters of words, with a certain concept or identifying characteristic.

Words young women use to describe their favorite fashion styles would include "creative," "flirty," "professional," "classic," "sexy," and others. Words that suggest that a person is particularly focused on the fit of clothes would include "cinches my waist," "shows off my legs," and "doesn't it fit great?" (Uploaded pictures and video, not only text, are part of the analysis.) Phrases that identify a person's interests, intentions, or needs would include "going to a party and need something to wear," "taking a vacation next week," "still paying off loans," "got a dog," and "getting married." Hobbies, job status, favorite television shows, books read—the social-media mine is filled with nuggets of every kind, each of them statistical grist for hints of behavior and buying habits. It is real people saying things in their own words—not a marketing focus group or a survey. And that self-revelatory data is the social currency of the millennial generation, not just freely, but eagerly offered. "They're actually giving you what their preferences are," Haydock observes, if you are skilled in listening to the data.

But that skill is not all technical. The words people use online are digital signals of behavior and buying habits, but only partial ones. They must be assembled, interpreted, and put into categories. What words or phrases fit into what category? How are words and categories related to each other? Numbers are attached to the words, categories, and relationships. Human communication, with all its ambiguity and nuance, is converted into an engineering artifact or, as Haydock says, "something I can reasonably apply math and computing to."

A seemingly unstructured resource is given structure. It is his team's "listening model," and the model varies by industry, product market, and demographic cluster. Veronica Vargas, a young IBM consultant, played a key role in shaping the listening model. Vargas brought retail experience and life experience as a millennial to Haydock's team. She worked as a buyer for the Dillard's department store chain for three years before she went to Columbia for an MBA, concentrating in marketing and statistics, and joined IBM in 2011.

In the Macy's project, Vargas worked alongside the quants. "They thought of me as a translator," she says. She helped in decoding the context and meaning of words and phrases. Not all messages or their authors were equal. The team's research, for example, identified more than 100 professional bloggers. It is a fair bet that Haydock is not familiar with the fashion blog *College Fashionista*, as Vargas is. She was not writing computer code, but as the project progressed, she recalls, "I really had to dig into the data."

The virtue of a small, diverse team, its members working side by side, is that judgment calls are made jointly by the business experts and the data scientists. Vargas takes a creator's pride for her part in making the listening engine, tuned for marketing to millennials, which will grow and get smarter. Yet being a participant in building data models has also given her an appreciation for their limitations. "There are no computer systems," she observes, "that are without human bias."

I n computing, a model is the equivalent of a metaphor, an explanatory simplification. It usefully distills, but it also somewhat distorts. Haydock is building models of human behavior. It is a different environment from modeling physical systems—Gallo's vineyard, GE industrial equipment, or even Nest's "thermal modeling in the home." Yet similar modeling techniques are employed. His work, Haydock explains, is physics with a twist. On projects, he takes the first cut at a software solution and he programs in Speakeasy, a numerical programming language developed for physicists at the Argonne National Laboratory.

But even complex physical systems, like the weather, operate according to physical laws of nature. Emotional human beings do not. Haydock acknowledges the challenge. "You're estimating stuff you can't see," he notes. Still, he is confident. "I can't see lifestyle directly in data," he says. "But I might be able to observe it as a data pattern with a strong and reliable correlation."

His task is similar to that of Wall Street quants who apply physics to financial markets. Haydock manipulates data to derive signals about human activity, just as quants use derivative instruments in finance. "It's the same as the science behind financial economics," he says. That is not necessarily a confidence-inspiring statement, given the lessons of the 2008 financial crisis. Quants didn't cause the crisis, but they played their part. Their risk models proved myopic because they were too simple-minded, unable to take account of the rich, chaotic tapestry of behavior, especially in times of stress.

Emanuel Derman, a physicist and former quant at Goldman Sachs, explained the perils of maths models in finance, in his 2011 book, *Models Behaving Badly*. After coming to Wall Street in 1985, Derman soon came to believe that physics models could be successfully applied to finance and economics—a belief he later abandoned. "In physics," he writes, "you're playing against God, and He doesn't change His laws very often. In finance you're playing with God's creatures, agents who value assets based on their ephemeral opinions." Modeling in finance, according to Derman, the director of the financial engineering program at Columbia, is by no means a waste of time. But, he cautions, "You have to understand what models are best used for, and then be very careful not to discard your common sense."

B ig mainstream companies, like Macy's, that are embracing data science are mostly doing so step-by-step. It is a competitive supplement to their business, not a replacement. Speaking at a retailing conference in early 2014, Terry Lundgren, chief executive of Macy's, discussed the joint work with IBM to "help us understand the millennial customer." Some of the "best learning," Lundgren said, came from using an old-fashioned technique. They put ten millennial-aged Macy's employees and ten IBM millennials together, he added, and listened to what they had to say about "how they choose to shop and what they look for in a retailer." The market researchers from the two companies wanted to tap the wisdom of the articulate few, in addition to the data signals from the social-media crowd. Data science is both new and promising, so, Lundgren told the audience, a willingness to experiment is part of the strategy. An immediate financial payoff, he said, is not the goal on "proof of concept" projects, which are calibrated bets that there will be a payoff down the road.

Haydock specializes in those kinds of exploratory commercial projects. He describes his essential skill as "torturing data," but he is also part social scientist. In the Macy's project, the goal was to reach new, younger customers. Macy's has new departments aimed at younger and older millennials—MStyleLab and Impulse. Haydock visited those departments in more than a dozen stores across the country, sitting as if he were waiting for a daughter, and observing. Time spent as a field anthropologist in stores in not uncommon. "There's a huge anthropological aspect to this," he explains. "It's about people and their behavior."

9

THE LONG GAME

The first time I visited Jeff Hammerbacher at the Mount Sinai medical center I got lost and a member of the administrative staff helped me out, pointing to a small office at the end of a row of work cubicles. That, she said, is where I would find "Dr. Hammerbacher," as if he were a physician. But his is no ordinary clinic. The place looks like a thriving Internet start-up. The walls are white boards, cluttered with equations. Rows of partitioned workspaces have large flat-screen monitors, mostly with the bitten-fruit logo of Apple. The shoulder-high partitions are a neon lime-green, an odd choice in décor, but not one that distracts the large room's young workers, who are intently hunched over their computer keyboards. This is operation central for a nerds' insurgency. The idea: put a bunch of quants inside a hospital, and see if they can foment a revolution in health care.

If the Mount Sinai campaign is successful in making medicine "an information game," as the scientist who recruited Hammerbacher put it, the health care hierarchy will be recast. The all-knowing doctor will no longer reign supreme. The data scientist will be there as well, monitoring and recommending treatments. Not the data scientist in person, of course, but the data scientist's handiwork—software algorithms bearing data-driven counsel. That is what Hammerbacher means when he speaks of turning medicine into "the land of the quants." And genomics, the digitally turbocharged study of the molecular building blocks of life, is the main vehicle pushing things in that direction. The human genome, in Hammerbacher's view, "represents the quantification of the core of what we are."

At Mount Sinai, Hammerbacher heads one team, among several, in a richly funded institute for what is called genomics and multiscale biology, which is the cutting edge of big-data medicine. The institute is only a few years old, but it is working on ambitious projects in the treatment of cancer, diabetes, Alzheimer's, and Crohn's disease. Hammerbacher has a particular interest in mental health and he would ultimately like to advance research in that field—his "long game," as he calls it. But medicine itself seems a long game, an industry steeped in rules and regulations, with approvals sought and required for everything from therapeutic practice to payment schedules.

Medicine may be the "best problem," as Hammerbacher puts it, but it is a thorny problem as well. That raises questions not only about how soon data science can have a real impact in medicine, but also about whether young data scientists like Hammerbacher will have the persistence for the long game. His career so far has been one of restless reinvention, changing course every few years. That is a virtue—the smart move—in the start-up culture of Silicon Valley. But tackling chronic diseases will take years of commitment. Hammerbacher wants to learn as much about medical science as he can, but the main place he can make a difference, he says, is in building the software tools to accelerate progress in data-driven medicine. "My tribe," he says, "is people who want to be major technology contributors in tools that will change the world." A data scientist, in good part, is a digital toolsmith with a larger vision. Hammerbacher will never win a Nobel Prize in medicine himself, but someone using the software his Mount Sinai team creates might someday.

H ammerbacher is fascinated by the workings of the complex machinery of the body. He describes disease as a biological system gone awry in some way. Often, he says, the best way to understand how something works is to learn what happens when it fails. He does that in computing. He studies technical reports on Internet breakdowns and outages, failures in highly complex networks, not unlike biological networks. Mental illness, in Hammerbacher's view, is "the most interesting way the central nervous system breaks." Discoveries and treatments, he believes, will result from "studying broken brains." Hammerbacher has some familiarity with broken brains, both through observation and his own experience.

In his first year at Harvard, Hammerbacher and another freshman, Steven Snyder, were close friends. Both were pitchers on the baseball team, both planned to major in maths, and both came from the Midwest (Snyder from Cleveland). Snyder had a quick sardonic wit, which Hammerbacher appreciated and found amusing. One day the two were walking through Harvard Square. Sitting on the sidewalk was a disheveled man with a sign, JESUS SAVES. Snyder put a few dollars in the man's cup, and told him, "Moses invests." Among his talents, Snyder was an accomplished amateur magician, and Hammerbacher recalls him performing for the patients in the cancer ward at Boston Children's Hospital. For Hammerbacher and Snyder, their freshman year was largely a shared experience. "We lost our virginity on the same night," Hammerbacher recalls, and quickly adds, "but not with the same girl."

Snyder flunked out of school at the end of his freshman year. After that, the two drifted apart. Snyder, under Harvard rules, had to stay away for at least a year. And Hammerbacher, who got bounced after skipping finals as a sophomore, was gone the following year. A few years later, when both were back at Harvard as juniors, Hammerbacher recalls visiting Snyder in his room. Finals were approaching and Snyder was behaving strangely. He had cleared out the furniture and was sitting on the floor, wearing only his boxer shorts, smoking marijuana. Something was wrong, but Hammerbacher and others thought Snyder was merely reacting badly to the strain of final exams. "We thought Steve really needed to chill out," Hammerbacher recalls, "but not that he was headed for a mental institution."

But he was. Snyder checked himself into Harvard's health center, which referred him to a psychiatric hospital nearby. Snyder was diagnosed as bipolar and was committed for several months. Still, after he returned from the hospital, Snyder's behavior became more erratic. Once, he jumped into the Charles River and swam for a distance before he was fished out; Snyder said he had wanted to swim to visit a friend—in Hawaii. Hammerbacher graduated in 2005. Snyder, who had been in and out of school, was back in the fall of 2006. One October morning, a despondent Snyder jumped out the ninth-floor window of a campus building. He survived that suicide attempt, but Snyder did take his life in 2008.

Snyder's death wasn't really a surprise, Hammerbacher says, but it hit him hard all the same. His sadness was tinged with remorse, since Hammerbacher's contact with Snyder had diminished to occasional calls and e-mail. By the spring of 2008, when Snyder died, Hammerbacher was living on the other coast, in Silicon Valley, working long hours at Facebook and soon to be off to found a start-up. His life was elsewhere, in motion and propelling ahead. Yet Snyder's death was personally unnerving. "Seeing Steve was my first glimpse of a full-bore mental health issue. And a part of me recognized that Steve and I were very much alike," he says. "I didn't address it then."

Two years later, in March 2010, Hammerbacher was forced to confront the problem, after rushing to the emergency ward and being admitted to a San Francisco hospital. He was having a raging panic attack that felt like cardiac arrest or a stroke: heart palpitations, sweaty palms, dimming vision, and a piercing pain inside the head. "You're very out of control of what your brain is doing," he explains. He was given Lorazepam, a powerful drug to treat anxiety disorders, and sent home.

Hammerbacher knew he had to change. His regimen was taking its toll. He calls it "living the start-up life." Perhaps, but Hammerbacher's version of it was well out on the self-destructive end of the scale. He often ate little, if at all, during the day. Instead, he was sustained by pounding back bottle after bottle of a potent, caffeine-packed energy drink, Hansen's Natural Energy Pro. Hammerbacher drank so much of the stuff that he ordered it directly from the bottler, by the case. Marathon workdays were capped off by nights of partying, drinking, and recreational drugs. A few hours of sleep, up the next day, and the routine started all over again. When he landed in the hospital, Hammerbacher was just twenty-seven years old. At the time, he recalled thinking of the 27 Club, so named for all the pop music icons that have died at twenty-seven, including Jimi Hendrix, Janis Joplin, Jim Morrison, and Kurt Cobain, from living hard, fast, and foolish.

His professional and personal life suffered. At work, Hammerbacher's appearance at meetings with customers became unpredictable, to the irritation of his colleagues at Cloudera. They passed it off as Hammerbacher's intensity, his absentmindedness, and his disregard for schedules. His behavior seemed rude and self-centered, but not a signal of personal distress. Yet it was more than Jeff being Jeff. "It was clear to me I was not keeping things together as well as I should have," he admits.

Hammerbacher's lifestyle was also threatening his relationship with Halle Tecco, the young woman who would become his wife. They met in January 2007, introduced by mutual friends in San Francisco, and her first impression was that he wasn't her type. He looked scruffy and sloppy, with hair disheveled, as if he'd slept on it. Tecco, by contrast, has always been "put together," in her phrase. When we had lunch at the Wayfare Tavern, a place of red brick walls and dark wood interiors in San Francisco, she wore a beige jacket, gray blouse, and black pants tucked in black boots. She's tall and slender, with shoulder-length brown hair. The look is professional, yet stylish. When she met Hammerbacher, Tecco was seeing a Brazilian who worked at Google, and was a DJ in his spare time, more her type, she thought. But Hammerbacher was persistent, and "he started growing on me" over long conversations, she recalls. Tecco and Hammerbacher shared similar backgrounds and upbringing, strivers from the Midwest. And the grooming gap was bridgeable. "I've cleaned him up a lot over the years," Tecco observes, smiling.

They are similar in ways but have contrasting temperaments that complement each other. When talking about his wife, Hammerbacher mentions, among other things, that she is polished, practical, and orderly, and has a personality with "not a lot of dark side." She is, he says, a source of "real ballast and balance for me." Where he often pays little attention to day-to-day things, like mealtimes and appointments, "I'm the opposite," she says. "Nothing falls through the cracks on my watch." As Hammerbacher's mother, Lenore, put it, "She's smart in ways he's not."

Toward the end of our lunch, Tecco offers a crisp summary of their differing personalities: "I wake up every day happy to be alive.

Jeff gets up every day and looks for a reason to be happy." That, she suggests, is the animating energy behind his restlessness, hard work, and penchant for reinventing himself every few years.

B y the time he was taken to the hospital, Hammerbacher and Tecco had been a couple for three years, but their future was in doubt. "I couldn't go on with him unless he changed," she says. Hammerbacher responded, and made some abrupt changes indeed. "I had to choose another path," he says. Gone were drugs and alcohol altogether, not even a glass of wine with dinner. "I just stopped," he explains. "It's a black-and-white thing for me." That alone wasn't a cure. He had a few more visits to the hospital in the following months with severe panic attacks. He went to a psychiatrist and was diagnosed with bipolar disorder and generalized anxiety, and he took mild doses of drugs for seven months. But he also found that sleeping six hours a night, eating reasonably regularly, exposure to natural sunlight, and exercise help a lot. Hammerbacher, by all accounts, is healthy and fit. "I'm very grateful that he turned around," Tecco says. "He really committed to being sober, and to his family, his friends, and his work." The drastic change in his habits, she says, appeared effortless—no 12-step programs or the like. "But I'm sure it was not as easy for him as he makes it look," Tecco notes.

Hammerbacher is managing an illness. Recognizing that reality has helped, giving him perspective and fostering empathy for others. "I've had troubles with mood and anxiety at times in my life," he observes. "That's a challenge for me. It's been very humbling." His experience and the demise of his friend Steven Snyder have taught Hammerbacher that people with mental illness are not suffering from a character weakness but are battling a disease. "I didn't understand that before," he explains.

Hammerbacher has become a champion of the medicinal power of data in understanding and treating mental illness. At a Silicon Valley conference in May 2011, Hammerbacher presented a short version of Steven Snyder's demise and his own bipolar diagnosis as the starting point in an appeal for data scientists to put mental illness on their radar. He cited numbers from the Global Burden of Disease Study, a collaboration of health experts worldwide, showing that mental health ranks third in terms of deaths and disability—behind infectious and parasitic diseases and cardiovascular disease. "It's huge, huge crisis from my perspective," Hammerbacher told the audience. In the United States, 46 percent of the population will develop a mental disorder over the course of their lives. The leading categories are anxiety, impulse control, mood swings, and substance abuse. Pointing to the 46 percent figure, Hammerbacher said, "It is one of those numbers that causes you to reevaluate what it means to be human and what it means to be sane and what it means to be crazy."

The opportunity, Hammerbacher said, is wide open. The causes of most mental disorders are little understood, and there are puzzling debates over the definitions, symptoms, and diagnosis. But vast, new data sets are emerging from genomics and brain-scanning imagery. The data will explode, but what will be in short supply is smart people to manage and make sense of the data. Mental illness, he suggested, is a data moon shot. "Think about mental disorders," he urged the audience, "as one of the most critical and most challenging data problems for us out there."

When Hammerbacher gave that Silicon Valley talk, he was already moving toward medicine as his next act. By then, he had become a director of Sage Bionetworks, the nonprofit organization to promote the sharing of medical data. In the fall of 2011, one of the nonprofit's founders, Eric Schadt, had moved to New York, taking a job at Mount Sinai. Shortly thereafter, Hammerbacher was on a business trip to the East Coast and dropped by Mount Sinai. A renowned researcher in genomics and biomathematics, Schadt had shuttled between the research labs of a couple of big pharmaceutical companies and biotech start-ups. Before Mount Sinai, he was the chief science officer at Pacific Biosciences, a gene sequencing company in Silicon Valley.

Schadt did not move east for an ordinary job. He was given the mandate to create an institute with a $100 million budget for the first five years, the Icahn Institute for Genomics and Multiscale Biology. Most of the funding comes from the man whose name is on the institute, Carl Icahn, the Wall Street financier and philanthropist. Icahn has taken a particular interest in genetics, having also paid for a genomics laboratory that bears his name at Princeton, his alma mater.

Hammerbacher came away impressed from that meeting with Schadt. He knew enough about health care to know that it can be a bureaucratic, slow-moving industry. "The default answer in health care is 'no,' but Eric was a person a lot of people were saying 'yes' to," Hammerbacher recalls. But at the time, he was in the exploration stage of his move into medicine. Besides, Hammerbacher says he rarely buys anything without looking at a comparable product. So he studied alternatives, meeting with scientists at Stanford, the University of Southern California, and the University of California at Santa Cruz.

Hammerbacher was looking for a commitment because he planned on making one himself. He was not seeking a leisurely academic perch, as a lecturer or adjunct professor. He wanted the funds and freedom to recruit a team, build technology, and have an impact. Mount Sinai promised all of that. And its leadership shared Hammerbacher's data-driven vision, none more so than Eric Schadt.

S chadt, like Hammerbacher, is a native Midwesterner. But unlike Hammerbacher, he has no fond memories of growing up there. He was raised in Stevensville, Michigan. His family, by his account, was poor, ultrareligious (evangelical Christian), and never valued education. "I hated Michigan every day," he says. After high school, he joined the air force, where aptitude tests showed an uncommon talent in maths. He was sent on a military scholarship to California Polytechnic State University, where he majored in applied maths and computer science. Later, he earned a master's degree in maths at the University of California at Davis, and a PhD in biomathematics at the University of California at Los Angeles.

Schadt is stocky yet energetic, walking the halls of Mount Sinai at a brisk pace. He has lumpy features and a heavy brow, with playful eyes and a head of thick brown hair. He would fit in a troupe of Shakespearean actors. Still, the most remarkable thing about his appearance if you've met him more than once, and not in the summer, is his wardrobe: a traditional white tennis shirt and a pair of hiking shorts. It's his uniform, all year round, even at industry conferences. His conversion to sartorial monotony, Schadt recalls, began one day when he was deep into postgraduate maths studies. "A deep math flow really does change you," he confides. "My mind really loves it." So while immersed in a maths reverie, it was time to get dressed for the day. Schadt went to his closet, opened the door, and was struck by a mundane epiphany: "Why waste my time thinking about this?" He grabbed the first things at hand and has stayed with those styles since.

At UCLA, Schadt chose biomathematics as a PhD as a way

to combine the abstractions of maths with the real-world study of biology. As he got into biology, though, he began to question. Biology, he explains, was "taught qualitatively as opposed to quantitatively. It seemed very naïve to me." That approach, he thought, led to a simplified understanding of disease. In molecular biology, the focus seemed to be on variants and mutations in single genes. As his study and research progressed over the years, Schadt concluded that human biology is a complex dance of many different networks—molecules, cells, tissue, organs, humans, and human communities—each interacting with the other. People with similar gene traits often have very different health outcomes. Chronic diseases like cancer, heart disease, and Alzheimer's are not caused by single genes, Schadt explains, but are "complex, networked disorders."

The "network of networks" perspective is why the name of Schadt's institute includes "multiscale biology." Science, he insists, is on the threshold of realizing the potential of genomics to deliver new discoveries in drugs, treatments, and personalized care. The advancing technology of genetic sequencing and data analysis, he adds, will drive success. The government-financed Human Genome Project, completed in 2003, cost $2.7 billion. Today, whole human genome sequencing, identifying all three billion chemical units in the human genetic instruction set, can be done for as little as $3,000. In a few years, Schadt predicts, the cost will be less than $1,000, and within five or ten years, less than $100, almost like a blood test today.

At Mount Sinai, Schadt and his researchers plan to combine genetic information with other data about patients such as weight, age, gender, vital signs, tobacco use, toxic exposure, and exercise routines. Each is a data ingredient, converted to numbers, in a sophisticated mathematical model used to test and predict how a person's genes, medical history, lifestyle, and environment interact to affect health outcomes—the likelihood that a person will develop a certain illness within three years, for example. That kind of large-scale measurement and modeling, Schadt observes, is fairly common in some other fields involving complex, dynamic systems like weather prediction or the physics of galaxy formation. "We're trying to move medicine in the direction of climatology and physics; disciplines that are far more advanced and mature quantitatively," he explains.

Beyond the mission and the ample funding, the appeal of Mount Sinai for Hammerbacher was the willingness of the heads of the medical school and hospital to let the data researchers be part of the hospital system. They would not be university researchers isolated from the delivery of care. Their work would be used to treat patients and they would have access to patient data, with appropriate safeguards for privacy. And a rich pool of patient data it could be. The Mount Sinai system, which has expanded in recent years, has seven campuses, 6,600 physicians, and millions of patient visits a year, representing every age, income, and ethnic group. The human melting pot of New York offers an unmatched diversity of patient data. "You go to Palo Alto and it's fantasyland by comparison, most of it anyway," Schadt notes.

The Mount Sinai medical system's chief executive, Dr. Kenneth Davis, and the dean of the medical school, Dr. Dennis Charney, spent most of their careers as researchers in psychiatry and neuroscience. Davis is an expert in Alzheimer's disease and schizophrenia, and Charney is a specialist in depression and resilience. Davis and Charney have been friends for three decades, and Charney explains that they are "researchers by temperament and training." The research mentality, he says, fosters a sense of looking for new things, experimenting, and taking risks. Charney is a dapper academic, heavyset with a trim beard, and longish hair combed straight back. When we met in his Mount Sinai office, he wore a crisp white shirt, beige suspenders, and a dark brown silk tie, beneath his white doctor's coat. Charney looks and acts like a man who would be equally at home in a medical lab or at a fund-raiser.

The big computing and data shop that Mount Sinai is setting up qualifies as a risk, and a costly one. Why do it, I ask him, when so much of the drift in thinking about the best ways to improve hospital care focuses on checklists and hand washing, that is, monitoring and standardizing the labor-intensive day-to-day tasks of doctors, nurses, and attendants that can cut down sharply on treatment errors and infections. Yes, Charney replies, quality-control programs to improve practices and processes are important. But health care, he says, needs a lot more. What's needed, according to Charney, is a dose of Silicon Valley. He speaks of the "digital mind-set," which he essentially defines as high-tech smarts combined with "we can change the world" optimism. Charney talks of a "paradigm shift," a term he and Schadt use frequently, and cites *The Structure of Scientific Revolutions* by Thomas Kuhn, a historian and philosopher of science, whose book in 1962 introduced the concept of the paradigm shift, when forces combine to produce a sharp break with the past. "We can break new ground and energize the biomedical research engine and medical education," Charney says, "by emulating some of the lessons from the digital revolution."

Hammerbacher found common ground with the Mount Sinai leadership. But he was also impressed that Schadt was persuading others with outstanding technology credentials to join him. One of them was Michael Linderman, a young computer scientist from Stanford. Linderman, Hammerbacher recalls, was "the first person I met who was there who I could sit down and write code with." Patricia Kovatch came to Mount Sinai after leading a team that built one of the world's fastest supercomputers, located at the Oak Ridge National Laboratory in Tennessee. When it was deployed in November 2009, the machine was the third fastest in the world, and it has been used to advance science by running simulations of the weather, earthquakes, and the birth of the universe. The giant machine was called Kraken, named after a legendary sea monster of Nordic myth. But after it was built, she recalls asking herself: "Do I just build a bigger machine or do I do something else?" She opted for the something else and decided to come to Mount Sinai. She is convinced, she says, that "this is the future of health care."

At Mount Sinai, she is building a supercomputer for genomics research. When I visited her in the fall of 2013, Kovatch's supercomputer was growing fast, from 9,000 processing cores to 20,000 by the end of the year. The machine can handle various tasks, but an important one is gene sequencing. At Mount Sinai, they are sequencing entire genomes—looking at all three billion nucleotides, the basic structural unit of DNA. Within that deluge of nucleotides, scientists have identified about ten million DNA segments called SNPs (pronounced snips), for single nucleotide polymorphisms, that have been linked to diseases in research studies. Consumer gene-testing services, like 23andMe, look at fewer than a million SNPs. At Mount Sinai, the ambitions are larger. They want to see the whole picture, the entire genome sequenced. To really advance research and treatments at Mount Sinai, it will have to do a lot of it, very quickly. The goal, Kovatch says, is to compress the time it takes from days down to an hour. She has named her supercomputer Minerva, for the Roman goddess of wisdom.

Hammerbacher joined Mount Sinai after Kovatch, so he followed her lead in naming his data-analysis computer cluster. He appreciates the mythology reference, having been a "mythology nerd" as a youngster when he read *Bulfinch's Mythology* several times. And he named his data-crunching cluster Demeter, the Greek goddess of the harvest. His system will be to harvest the data and information from across the hospital and to try to convert it to knowledge. Some of what Demeter can harvest is data generated by Minerva. A supercomputer is a processing speed demon, while a data-analysis cluster is almost contemplative by comparison. Demeter gathers data, and its software algorithms go hunting for patterns of illumination.

I t wasn't until 2013 that Hammerbacher began spending most of his time in New York. By the fall, he had built the first couple of stages of the Demeter cluster and had begun to recruit a team that he hoped would grow to about ten people. On a snowy Saturday in December, Hammerbacher, his first four recruits, and a couple of others gathered for an all-day meeting. The session was held in the downtown office of a boutique venture capital firm, headed by a former Harvard classmate of Hammerbacher's, who let him use the space for the day. Seven young men sit at a round table in a glassed-off conference room. Hammerbacher calls it a "level-setting" meeting, "so we're all on the same page."

Three of the four new hires had not yet started at Mount Sinai. So with his slides displayed on a large flat-screen television monitor on the wall, Hammerbacher runs through everything from the history of the hospital (begun as the forty-five-bed Jews' Hospital in the 1850s) to the various computer systems and data sources available at Mount Sinai today: electronic health records, bedside telemetry, genomics, diagnostic images, lab test results, discharge and billing data, and a biobank with plasma and DNA samples from nearly 30,000 volunteers.

True to form, Hammerbacher informs the group that under his leadership meetings will be few: once a week to discuss current progress, and once a month to talk direction and vision. The presentation on the hardware and software of the Demeter computer cluster is detailed and acronym-filled. There is also an introduction to the intricacies of medical data—for example, different shorthand codes (of letters and numbers) used by different hospitals, software companies, and medical standards organizations. For a single disease, there may be more than twenty different codes. To make sense of the data, Hammerbacher's team will have to write software to automatically sort through the medical-coding Babel. "You thought sequencing was tricky," Hammerbacher observes. Throughout the day, as he surveys projects and opportunities, Hammerbacher offers similar asides. "As you can see, this is not the sort of thing you can automate trivially." "You realize these are really messy problems." "Welcome to biology." At one point, one of the new recruits chimes in, "This is going to be interesting."

The "level-setting" gathering on that wintry December day points to an underlying issue not only for the Mount Sinai endeavor but also for the future of data science in general. In broad strokes, the direction is irrefutable and the vision is clear: data and the smart software tools that turn raw data into knowledge will increasingly fuel discovery and decision making. But how quickly and where can progress be made in business and science?

Vision has rarely been the problem in computing. The basic ideas in the field of artificial intelligence—the technology that makes big data smart—date back to the 1950s and before. The term "artificial intelligence" was coined in 1955, and the theory of computer-simulated intelligence was set in 1937. That was the year that Alan Turing, a British mathematician, computing pioneer, and famed wartime codebreaker, published a paper in which he described what he called a "universal machine"—a theoretical computer. He started by demonstrating that a clerk, given the proper instructions and limitless supplies of paper and time, could solve any problem that an expert mathematician could answer. The implication was that a universal machine, a computer, could simulate work of any computing device, including the human brain. After Turing, all the rest of computer science can be glibly seen as just engineering.

Make the vision happen. That is the challenge at Mount Sinai, and for so many ambitious big-data initiatives. There is some very promising work already at Mount Sinai. One program is for personalized cancer therapies. It involves the genetic sequencing of a patient's healthy cells and then sequencing the patient's cancer tumor. The misbehaving gene network is identified, analyzed, and targeted with custom-designed drug therapies. The hospital researchers also have genomic programs for Alzheimer's disease, Crohn's disease, and diabetes. Their projects have attracted funding from the government's National Institutes of Health and private companies. There is plenty of work to be done, and progress to be made, Hammerbacher tells his group. "Embedded within these clinical problems are genome data management problems," he says. "You can't solve one without the other."

Hammerbacher is also intrigued by a very different kind of health care menace: hospital-acquired infections. The idea would be to use swabs, sensors, and data analysis to combat such infections. Nationally, hospital infections are a huge problem measured in suffering and in dollars, causing or contributing to nearly 100,000 deaths a year and costing an estimated $10 billion annually. The goal, Hammerbacher explains, would be to track the outbreak and spread of such infections, and apply data analysis to predict where the risk is greatest, thus targeting treatment and prevention measures. Curbing hospital-acquired infections, he admits, may lack the glamour of hunting for personalized cancer cures. "But if we want to affect the hospital and patient outcomes in the next year or two, this area is very promising," he tells the team. "This one I've been holding in my back pocket."

In the days after the meeting, I talk to a few of the recent recruits to Hammerbacher's team, all in their late twenties to early thirties.

Alex Rubinsteyn had just completed his PhD in computer science at New York University's Courant Institute and job offers flowed in. An earlier stint working for a Wall Street trading firm convinced him to avoid finance, so much so that he participated in the Occupy Wall Street protests. Rubinsteyn was mostly interested in start-ups and he almost joined a data start-up in San Francisco, but its founder made the mistake of introducing him to Hammerbacher. A short conversation turned into a long one, and Rubinsteyn came away, he says, with "a clarified sense that I needed to work on something that mattered." The work at Mount Sinai, he adds, seemed "significantly more important and useful and ethical than anything else I was going to do."

None of the team members are making a financial sacrifice in any normal sense. They are well paid. But they are salaried employees at a nonprofit medical center. It's not a start-up with hopes of cashing in or a big technology company doling out stock options. They came for the work. Tim O'Donnell was doing broadly similar research as a computer scientist at D. E. Shaw Research, which specializes in computational biochemistry. But he was lured by the broad opportunity at Mount Sinai to pretty much pick and choose projects in disease discovery and treatment. Trying to use data as a lever to change hospital care, he says, is "a good, ambitious goal," but likely a long-term one. O'Donnell may well be gone to a start-up before that happens. He figures he will stay at Mount Sinai for two or three years. In that time, he hopes to contribute useful software for medical research, work that is "ultimately helping patients."

How far Hammerbacher gets, or goes, at Mount Sinai is uncertain, whether it becomes his "long game" or not. But to him, it feels like a very good place to apply his quantitative skills to a very human need, health care. The right balance, you might say.

In fact, "balance" is a word he often uses when he's talking about himself.

For Hammerbacher, the quant side is always there; it's not just a skill, but also an instinct and a philosophy. But he's not all cool logic and life by the numbers. He has his emotional, intuitive side. His life decisions and habits increasingly reflect the two, as if a search for balance. It requires conscious choice and vigilance. "It takes discipline to keep myself in balance," he says. Take Hammerbacher's practice of reading two books in tandem, one technical book and one work of history or fiction. The first, he says, feeds his "numerical imagination"—the story the numbers tell you. The second nourishes his "narrative imagination"—the story told in people's lives, the experience of events in history, and the mystery of human relationships.

"Just because you can't measure something easily doesn't mean it's not important," the data scientist concedes.

10

THE PRYING EYES OF
BIG DATA

The Kodak camera, introduced in 1888, was a masterstroke of technical innovation that soon became a hit product. It was costly at first—$25 was a lot in those days—but the price would decline, down to $1 for the Kodak Brownie in 1900. Yet even at its initial price, the Kodak camera cost less than the hefty, wet-plate cameras of the time, which rested on tripods. But a larger difference was that the Kodak was a film camera. It could quickly capture images, recording people in normal poses and spontaneous moments. The wet-plate cameras required their subjects to present themselves in frozen stillness, while Kodak introduced the term "snapshot" into the vernacular. And the Kodak camera was easy to use, a consumer product rather than a chemistry project. Its catchy advertising slogan: "You press the button, we do the rest."

Still, the popular new image-capture technology had its drawbacks. A new breed of social misfit surfaced, the Kodak-wielding "camera fiend," mainly seen at seaside resorts snapping pictures of young women bathers. Freewheeling picture taking was sometimes unwelcome in other public spaces. Kodak cameras were briefly banned at the Washington Monument, according to historian and author David Lindsay. In an editorial, the *Hartford Courant* bemoaned the loss of privacy as Kodak cameras multiplied. "The sedate citizen," the Connecticut newspaper declared, "can't indulge in any hilariousness without the risk of being caught in the act and having his photograph passed around among his Sunday School children."

More than a half-century later, in the 1960s, a very different technological advance challenged common notions of privacy—the mainframe computer. That's when the federal government started putting tax returns into the giant computers, and consumer credit bureaus began assembling databases containing the personal financial information on millions of Americans. Many people feared that the new computerized databanks would be put in the service of an intrusive corporate or government Big Brother. The author Vance Packard made a career writing deftly timed assaults on social ills like manipulative advertising (*The Hidden Persuaders*, 1957) and planned obsolescence (*The Waste Makers*, 1960). In 1964, Packard took on the threat to privacy posed by the proliferation of computerized databases of personal information and new surveillance techniques, in *The Naked Society*. Packard referred to such methods as "the hidden eyes of business." The privacy scare at the dawn of the computer age "really freaked people out," observes Daniel Weitzner, a former senior technology policy official in the Obama administration, who is director of the MIT Information Policy Project. "The people who cared about privacy were every bit as worried as we are now."

Today, our concept of privacy is under threat once again—this time by the technologies of big data. The response, as in the past, will likely be a step-by-step evolution that involves changing some attitudes and changing some rules. As those Kodak cameras spread, for example, people became more comfortable having them around and with snapshot picture-taking in public. In certain settings, privacy was redefined, and lost. But the mass-market appeal of the technology prevailed.

The mainframe-era challenge to privacy led to legislation. In the United States, the main laws passed were the Truth in Lending Act in 1968, the Fair Credit Reporting Act in 1970, and the Privacy Act of 1974. Together, the laws set standards for the collection and use of personal data and helped lay the foundation for a flourishing consumer economy, fueled by credit. There was fierce opposition from industry at first, but the legislation proved to be in everyone's self-interest—new rules of the road for the big-data technology of its day.

A similar pattern of public policy and personal adaptation is the best bet for how the tension between privacy expectations and big-data technology will play out over the next several years. "We're working our way not to a solution, but to an accommodation with this technology," says Edward Felten, a computer scientist and former chief technologist in the Federal Trade Commission, the agency mainly responsible for safeguarding the privacy of Americans.

Felten has a spacious office at Princeton, where he is a professor of computer science and public affairs and the director of the Center for Information Technology Policy. Tall and lanky, with graying hair and rimless glasses, Felten has been applying his computer science skills to matters of public concern for years. He is an expert in computer security, and he has uncovered the software vulnerabilities in products ranging from widely used programming languages to the music industry's digital locks to the code in electronic voting machines. He was a witness for the Justice Department in its landmark antitrust case against Microsoft. Felten also testified on behalf of the American Civil Liberties Union in its suit challenging the National Security Agency's collection of the telephone call records of American citizens—the surveillance program disclosed by Edward Snowden's leaks—as a violation of the Fourth Amendment's protection against unreasonable search and seizure.

In a lengthy conversation, I happened to bring up the old Kodak experience. Felten reminded me that we revisited that issue about a decade ago, when fairly high-quality cameras became a standard feature on cell phones. Soon, shots surreptitiously taken in locker rooms and public showers were being posted on the Internet. Privacy advocates protested, and a few municipalities passed bans on cell phones in public bathrooms, locker rooms, and showers. Some cell phone makers responded by making the click sound louder when taking a picture, so anyone nearby could hear.

But mostly, behavior changed. People who took the pictures were not seen as clever but as dolts. "What emerged was a pretty good set of social conventions about when it's okay to take a cell phone picture and when it's not," Felten says. Not everyone would agree. Yet the embarrassing photos and videos posted on Facebook and elsewhere are mainly party shots where there is little reasonable expectation of privacy today. Privacy groups and public officials aren't advocating curbs on smartphone picture taking.

A social prohibition is an informal agreement, and there are always a few bad actors, but for the most part that is how things sorted out with mobile phone photos. "We could do that," he adds, "because the activity itself is relatively in the open."

Modern digital data collection, however, is mainly a hidden activity. The algorithms used to parse the data for patterns are typically proprietary—software black boxes. And the technology has raced ahead of the old rules and definitions. Traditional online privacy standards are an artifact of the 1990s, intended as a light-touch policy to nurture the development of an emerging industry. The key concept was "notice and choice." Websites would post notices of their privacy policies and users could then make choices about the sites they frequent or not, based on their privacy preferences. But few people read privacy notices, cloaked in legalese, and the notices can't hope to capture the intricacies of today's data economy. "There is a fundamental gap," Felten says. "Consumers don't know what is happening, so they can't make informed decisions."

As we talked, Felten raised several concerns about navigating through what he called "this increasingly observed and classified world." One of his themes was how big-data technology is outrunning public understanding and policy, as in the failure of the old "notice and choice" approach. Another example, he notes, is the definition of the data that identifies a person. Until recently, that seemed straightforward. It was information that identified you or was directly linked to you—your name, Social Security number, phone number, credit card numbers, and bank accounts, for example. This is known as "personally identifying information," and privacy regulations have focused on that definition of personally sensitive information.

Yet today, we live in a mosaic of connected data—much of it self-revealed on Facebook, Twitter, and elsewhere online—so that weak signals become stronger and stronger when combined to create a "social signature" of sufficient detail that it effectively identifies an individual. All of that data can even be used to work backward to come up with directly identifying information. Using information people post on social networks and other publicly available data, Alessandro Acquisti and Ralph Gross, computer scientists at Carnegie Mellon, were able to accurately predict the full, nine-digit Social Security numbers for 8.5 percent of the people born in the United States between 1989 and 2003—nearly five million Americans. It was a research project, but Social Security numbers are magic keys to identity thieves and other crooks.

Businesses, Felten points out, have the ability to pinpoint characteristics about individuals without violating any legal prohibitions on the use of personal information. "We need to get beyond this concept of personally identifying information," Felten says, "because the rest is deemed by default to be harmless." Through the inferential engines of big data, companies can often accurately predict if a person has a chronic disease or is financially strapped. Felten compares the current state of affairs to the digital equivalent of attending a conference with name badges. But instead of names, the people are wearing badges that say, "I'm a diabetic" or "I'm deeply in debt." "That's not considered personally identifiable information," he observes. "But it's much more sensitive information than your name."

A Senate staff report in December 2013 offered a glimpse of what modern data brokers offer their corporate clients. These companies, like Acxiom, Epsilon, and Experian, compile extensive dossiers on millions of individuals and families, tapping data sources that include public records, consumer purchases in physical and online stores, and Web browsing histories. The Senate staff got limited cooperation from the data brokers, but their documents did show how they group consumers into economic clusters from the affluent ("Power Couples" and "American Royalty") to the financially vulnerable ("Burdened by Debt: Singles" and "Fragile Families").

At times, the massive data collection and analysis can deliver pinpoint accuracy, unnervingly so. In January 2014, Mike Seay got a marketing pitch in the mail from OfficeMax, the office supplier. Seay, a resident of Lindenhurst, Illinois, was an occasional customer of OfficeMax and got mailings from the retailer every so often. But this one was different. Below his name, the second line of the letter read, "Daughter Killed in Car Crash." His seventeen-year-old daughter, Ashley, had been killed in an auto accident the year before. Appalled, Seay posted the letter on Facebook, and a local television reporter, Nesita Kwan from Chicago's NBC5 station, soon saw the letter and a story. When he was interviewed on the local news program, Seay posed some pointed questions, "Why would they have that type of information? Why would they need that? . . . And how much other types of information do they have if they have that on me, or anyone else?" His initial complaint to OfficeMax was met with disbelief. But once the story got out the company apologized and explained that the mailing resulted from a list provided by an unnamed supplier.

The OfficeMax episode was obviously a miscue. But it and other similar cases offer occasional, inadvertent peeks inside the curtain of big-data marketing and how it can be so detailed and accurate that it becomes disturbing.

T he largest of the data brokers is Acxiom, headquartered in Little Rock, Arkansas. The company has collected data on hundreds of millions of consumers worldwide. For each person, the company boasts, it typically has thousands of signals of behavior, or "attributes," culled from public records, purchase data, and online browsing habits. The public and inferred information ranges from age, race, and sex to political affiliation, vacation hopes, and health concerns. Put it all together, and Acxiom has a deeper data view of most American adults than any federal agency or Internet company. For that reason, Acxiom is a leading supplier of consumer information to the corporate world, and it is the bane of privacy advocates. Jeffrey Chester, executive director of the Center for Digital Democracy, calls Acxiom "Big Brother in Arkansas."

But in the summer of 2013, Acxiom opened its data vault a bit to let people see what information it had about them. It was done at a time that the data broker industry was under scrutiny both by the Federal Trade Commission and Congress. The initiative was an olive branch from Scott Howe, who took over as Acxiom's chief executive the previous year, joining from Microsoft, where he was a senior executive in its online advertising group. Howe sees Acxiom not as a data miner but as a modern "data refinery" that takes the noise out of data signals to make marketing more effective for companies and more useful for consumers.

At Howe's direction, Acxiom set up a website, AboutTheData. com. Visitors to the site can put in their full name, address, birth date, and last four digits of their Social Security number, and see the "core" data Acxiom has about them. An FTC commissioner praised the Acxiom site as a step toward greater openness and transparency by data brokers. Privacy advocates criticized Acxiom for disclosing only selective facts and little of the deeper analysis the company markets to corporate clients, placing people and households into categories like "potential inheritor," "adult with senior parent," and "diabetic focus."

In an article about Acxiom and its new website, my *Times* colleague Natasha Singer observed, "Visitors who log in may be surprised at the volume of information that may be available and the detailed picture it can give of their personal lives." As an example, she went through the site's results for Howe himself in the category of "household interests." Then, she asked Howe about his interests, as identified by Acxiom, and what might have prompted its data-driven deduction. In his case, the data-generated associations were remarkably precise: "health and medical (he subscribes to health industry trade journals and founded a site called

Health123.com); crafts (he periodically works with stained glass); woodworking (he paid for undergraduate education at Princeton in part by working as an apprentice carpenter); tennis (he was on his high school team); gardening (his wife subscribes to Fine Gardening magazine); and religious/inspirational." Howe is a churchgoing Methodist.

That sort of profile is pretty impressive. But Acxiom is not always so astute. In early 2014, I decided to see what the data refiner knew about me. I typed in all of the requested personal information and clicked on the "submit" button. The first category that appeared was general characteristics. Acxiom's database correctly identified me as a male, a Caucasian/white, a registered Democrat who votes regularly, and a professional/technical worker. But it reported that my marital status was single. That was news to me, and to my wife of more than thirty years. The section with household economic data listed our household income at about half of my salary.

Elsewhere, Acxiom's algorithms had determined that December is the month when I renew my car insurance. For many years, as a foreign correspondent for the *Times* and then returning to live in New York, I have not owned a car or purchased an auto insurance policy. The household interests section that Acxiom produced for me, with twenty-three entries, looked to be a fairly random hodgepodge. It seemed that the list could mostly be explained by finding the name of any magazine we ever subscribed to. Interests in "cooking" and "gourmet cooking"? Not a lot of that goes on in our household. But if you look at all of the shelter magazines that have been delivered to our apartment over the years, and that my wife has written for *Martha Stewart Living* among others, it would be an understandable inference to make. Some were more puzzling, though. One listed interest was "Christian families." The only explanation I could think of is that my daughter attended an Episcopal-affiliated school from nursery school through eighth grade, St. Hilda's and St. Hugh's School. But if I were single, as Acxiom's data assumed, why would I have an interest in Christian families? Our household interest in "hunting/shooting" was another puzzler. Last year, I did purchase a jacket from Filson, the Seattle maker of outdoor clothes. Perhaps that could explain it, since the last time I shot a gun was in the late 1980s, outside Helsinki, at a skeet-shooting range with a Finnish cabinet minister.

Mostly bemused, I made a few changes in Acxiom's data profile of me, correcting the marital status and income. It's tempting to just make fun of the outcome. I wasn't surprised by the accuracy of the data broker's view of me; I was surprised by how clueless it was. Big data is a dummy, after all. Yet, if a data gatherer is so wrong about the basics, what happens when the information is linked with other data to make predictive models of my interests and behavior?

S ome data-driven mistakes are relatively harmless, others less so. If I get a junk-mail pitch from the National Rifle Association because a data broker has flagged me as a shooting enthusiast, so what? Having received a couple, I don't mind. The real issue, though, is discrimination by statistical inference. In the online economy, the news and information you see and the commercial offers you are presented are increasingly determined by the assumptions a computer algorithm makes about you, right or wrong. If the latter, Felten notes, "your decision space is reduced accordingly because they get you wrong."

The larger danger is if groups of people are systematically either misclassified or discriminated against. The Obama administration's big-data report, published in May 2014, featured that warning. The report praised the current benefits and future potential of big data but cautioned that the same technology, if used improperly, has "the potential to eclipse long-standing civil rights protections in how personal information is used in housing, credit, employment, health, education and the marketplace."

Latanya Sweeney, a computer scientist at Harvard, published a research paper in 2013 that offered an example of the problem. The delivery of online advertising, Sweeney explained, is a "socio-technical construct," and her study combined sociology and computer science. She did online searches using more than 2,100 names that previous research had divided into two groups: first names predominantly assigned at birth to black babies, like DeShawn, Darnell, and Keisha; and first names overwhelmingly given to white babies, like Geoffrey, Matthew, and Emma.

When searching for names identified as black, the ads delivered by Google's AdSense program were far more likely to include results that suggested an arrest record. So a search for "Trevon Jones" generated a text ad, "Trevon Jones, Arrested?" A neutral ad for the same personal search would be, "We Found: Trevon Jones." For the research, the searches were done on the Google website and a popular news site, reuters.com. A search for a black-identifying name on google.com was 25 percent more likely to produce an ad suggestive of an arrest record.

The suggestive ads appeared regardless of whether the websites being advertised, which supply information and background checks on individuals, have a record of an arrest on a person. Sweeney, a computer scientist at Harvard, who is black, searched her own name, "Latanya Sweeney," and for "Latanya Lockett," both of which generated ads with the word "arrest." After clicking on the ad link and paying a subscription fee, the background check site had no arrest record for "Latanya Sweeney" but it did for "Latanya Lockett."

Sweeney's paper describes the potential harm of such suggestive ads, since searching online for information about people is routine. "Perhaps," Sweeney begins, "you are in competition for an award, an appointment, a promotion, or a new job, or maybe you are in a position of trust, such as a professor, a physician, a banker, a judge, a manager, or a volunteer." Or, she continues, you want to join a club, make new friends, or date someone. Alongside the search results about your accomplishments is an ad implying that you may have a criminal record. "Worse," she writes, "the ads don't appear for your competitors."

This doesn't qualify as job discrimination, unless you could prove that an online ad prevented you from getting a job or promotion, which would be practically impossible. The ads themselves may well be deemed commercial free speech. Still, in a world of Internet-connected data and algorithms, real discrimination can be the effect, even if it is unintentional. "Technologists," Sweeney concludes, "may now have to think about societal consequences like structural racism in the technology they design."

The algorithm makers did make changes in the case studied by Sweeney, a well-known privacy researcher, who in 2014 became the Federal Trade Commission's chief technologist. Google no longer presents links to ads alongside the search results for the names of people. In Sweeney's research, the ads with "arrest" in the links came from a background-checking site, instantcheckmate.com. In early 2014, I searched some of the names in Sweeney's paper on the reuters.com site. Ads with links to sites offering public records were delivered, including ones from instantcheckmate.com. But the word "arrest" did not appear. A search for "Latanya Lockett" delivered a link to instantcheckmate.com that began, "We Found: Latanya Lockett." A search for "Latanya Sweeney" brought a more general come-on, "Looking for people in the US?"

So much of data science is, in Sweeney's phrase, a "socio-technical construct." And her research points to the importance of human judgment as a vital check-and-balance on this technology. Just because an algorithm finds a correlation, that doesn't necessarily mean you should exploit it. Given the nature of American society, it is probable that people with black-identifying names are more likely to have been arrested, on average, than others. Yet when an algorithm makes that assumption about people in that group, it is big-data racial profiling—again, discrimination by statistical inference. "Structural racism," in Sweeney's phrase, is a bad thing.

B ut the situation is not as clear-cut with other forms of discrimination. A market economy, by design, is a finely tuned engine of discrimination. Companies are constantly seeking out the best, most profitable customers. And consumers search for the best prices and goods and services that they value most. This is the good discrimination of seeing differences and allocating time, energy, and money accordingly. Modern marketing is about clustering consumers into smaller and smaller groups. And big-data techniques, as we've seen, only accelerate that trend by knowing more and inferring more about ever-smaller groups, even individuals. Yet the line that separates market segmentation and personalization from discrimination is a fine one.

What to do with data-driven insights is increasingly going to be a judgment call that corporate managers will have to make, based on their notions of ethics and self-interest. Take a hypothetical but plausible example: a company's call-center reports, customer data, and social-media tracking show that single Asian, black, and Hispanic women with urban zip codes are most likely to complain about the quality of products and service. But Asian women whose complaints are resolved become some of your most valuable customers. Should Asians get preferred treatment over black and Hispanic women when resolving complaints? Or a health insurer is seeking to grow by enrolling previously uninsured Americans. Under the Obamacare legislation, an insurer cannot discriminate against people with preexisting conditions. But applying data science to health blogs, browsing habits, social media, and profiles from data brokers can identify people most likely to have diabetes or depression. Do you systematically exclude those people from your marketing?

Discrimination, as a legal concept, focuses on the treatment of people in groups, by ethnicity, gender, or age. Big-data methods make it possible to assemble people by interests and characteristics that are far more detailed than traditional demographics. The technology also affords the opportunity to discriminate in another way—among people down to the individual level. That is what Michael Haydock, the IBM data scientist, was talking about when he says he has the ability to be "as creepy as you can possibly imagine." To explain, he says skilled data scientists can now accurately "infer a disease state"—like diabetes or depression. Should a pharmacy chain, for example, target disease-related offers to people the data says have those conditions? Haydock himself would advise against it, unless the person volunteers the information. Otherwise, the company, he says, has "violated a personal relationship" by using sensitive information "that's between an individual and his or her doctor."

As big-data technology advances, corporate executives are going to have to make judgments about what kinds of discrimination they will and will not allow. For companies, privacy will become less an issue to be managed when some breach occurs than a part of a company's brand. Will people download the company's smartphone application? Will they sign up for store credit cards? Both are conveniences, but both are also tracking devices. If they trust the company, convenience will trump privacy concerns. Marketers have a lot at stake. A greater measure of openness and disclosure on the part of companies, almost surely, will have to be part of the answer instead of treating their data technology as a trade secret and a black art, as most do now. Michael Schrage, a research fellow at the MIT Center for Digital Business, put it succinctly. A company's algorithms and data-handling practices, he says, should be "both fairly transparent and transparently fair."

After seeing how little Acxiom knew about me from the data it had collected, I decided to give corporate America another chance. I asked IBM what it could find out about me. Its response was to steer me to Michelle Zhou, a computer scientist at the IBM Almaden Research Center in San Jose at the time. She was leading a team that is developing "hyper-personalized systems" designed to decipher a person's personality traits, values, and needs. Most efforts to mine digital footprints for clues about people typically look at a person's browsing habits, social network, and the subject matter of online messages. But Zhou's project—a software program called KnowMe—analyzes the language choices people make in their Twitter posts. It combines artificial intelligence with psycholinguistics to identify personality traits. For Zhou, Twitter posts have two virtues—they are accessible to researchers and they are informal expression. "It's writing that represents who you are," she explains, a medium in which it is "harder to mask the traits."

Zhou is a diminutive dynamo, with a modified Louise Brooks bob. She wears black-frame glasses with wooden bows and, on the day I first met her, a black jacket, maroon skirt, and leather half boots. She grew up in Chongqing, a major city in southwest China, home to twenty-nine million people today.

Both her parents were physicians. The family had relatives and friends scattered across the United States, and one of them recommended Michigan State University. Zhou got her master's degree in Michigan, before heading to New York and Columbia, where she earned her PhD. At Columbia, she met another student who would become her husband, Bill Yoshimi, a Japanese American from Nebraska. When she worked at IBM's Watson lab in suburban New York and later at an IBM lab in Beijing, he worked for Goldman Sachs, applying his data science skills to risk management. When she transferred to Silicon Valley, Yoshimi got a job at Google. "My husband's great," Zhou says, with characteristic verve. "He's been very flexible." She jokes about her own personality, far out on the extrovert scale—no Twitter-word analysis required.

Much of her research career has focused on applying artificial intelligence to make it easier for humans to communicate with computers—a branch of a discipline called "human-computer interaction." Her work has combined computer science and psychology, and involves the computer trying to figure out what the human being wants. So software to deduce a person's personality like KnowMe, Zhou says, is a natural, next step. It builds on statistical methods for measuring personality that have their theoretical origins in the nineteenth century. But the assessment techniques have really become rigorous and refined in recent decades, with their accuracy verified by standardized tests, to determine a person's personality type, basic values, and human needs.

The Big Five personality traits, for example, are: openness to experience, conscientiousness, extraversion, agreeableness, and neuroticism. The basic values are: self-transcendence, conservation, self-enhancement, openness to change, and hedonism. The needs are: ideals, harmony, closeness, self-expression, excitement, and curiosity. Each term has its definition, and the strength or weakness of each is measured by psychometric tests.

Zhou's group also relied heavily on research in the field of psycholinguistics, like that done by James Pennebaker and Tal Yarkoni, both now at the University of Texas at Austin. Certain words, research shows, are correlated with personality traits. Some of the associations might seem obvious, but others are less so. Words that suggest openness, for example, include: folk, of, humans, narrative, films, decades, and blues. Word signals for extraversion: drinks, restaurant, dancing, crowd, glorious, and pool. And for neuroticism: awful, though, worse, sort, lazy, stressful, and ban.

The goal for Zhou has been to design an engine for making these personality-revealing connections, out in the real world. To be commercially practical, the entire process has to be computerized and unseen by consumers. "You can't very well say, Welcome to our store, and would you like to take a personality test?" she observes. Zhou and her team determined that they needed at least 200 tweets from a person to provide enough words of enough variety to analyze reliably. Then, they found 256 people—IBM employees—who had produced at least 200 tweets, were willing to have them be studied for the research project, and were also willing to take standard personality tests. The standard tests became the "ground truth," the yardstick by which the results of the computer analysis would be judged. For 81 percent of the twittering subjects, KnowMe pretty much matched the results of their formal tests for personality type, basic values, and needs.

I didn't take the standardized test, but Zhou and KnowMe went to work on my tweets. This time, as with Acxiom, the outcome was amusing—but for an entirely different reason. It was uncanny how on target the program was, almost as if I were looking in a personality mirror. That was even more striking because of the way I use Twitter. I am not a particularly active tweeter, and I don't put my life out there. Anyone who follows me is not going to learn what I had for breakfast or other incidental details of my life. When I tweet, it is usually about things I've read that I find illuminating or amusing. And, in the interest of full disclosure, I do usually tweet my own articles these days. The personality scorecard Zhou sent me was essentially a color wheel with traits and characteristics identified by different colors and sizes. Alongside each was a percentage, or a grade of "high" or "low." There were a few scores that seemed off target or contradictory. If I am low (18 percent) on self-transcendence, defined as "showing concern for the welfare and interests of others," why am I fairly high (67 percent) on altruism and high (94 percent) on sympathy? Still, the program mostly got it right. I scored high in openness to experience and curiosity. I would hope so. I'd definitely be in the wrong profession if I scored low on those two; I've often said the appeal of being journalist is that it gives you curiosity license.

I scored low on excitement, which is defined as "upbeat emotions, and having fun." My wife and daughter loved that result. They never tire of telling me that I'm "totally lacking in enthusiasm." I like to think I'm even-tempered and have a dry sense of wit. But score that one to KnowMe. I did take some comfort from having scored high in closeness, which is defined as "being close to family and setting up home," and having scored low in traits like anger and anxiety. It didn't seem that I was "lone nut with a gun" material, which is what I most feared in this exercise, knowing I would write about the outcome, regardless.

At a technical conference in the spring of 2014, Zhou presented a paper, written with two colleagues, which described the KnowMe system. But IBM's sights are really on a product rather than a research paper. So another link in the chain of inferences is needed. Automating the link between words and traits is an achievement, but the payoff comes from linking traits to behavior and buying habits. Help on that front comes from an established cottage industry that uses techniques like "personality mapping" to determine the intrinsic traits and needs that prompt people to buy things. "When people make choices, they have different constraints, different limitations," Zhou explains. They value things differently. For example, people with high scores for closeness (homebodies) are good prospects for home decorating products, home and gardening magazines, and boxed sets of period television sitcoms, like *That '70s Show*. Those who score high on ideals (perfectionists) are inclined to buy organic food and pet clothing.

There are all sorts of potential uses for IBM's personality identifying technology, if they can scale it up. Personally, I would like to see it become a consumer service. Let people see themselves more deeply, as revealed by their own tweets and analyzed by some clever software. IBM is not a consumer product company, but it could be the technology engine, partner, or licensor to a hot new start-up. The technology could become a lucrative arm of the quantified-self movement. Why not go beyond health-monitoring wristbands and smartphone applications? Democratize personality measurement.

Zhou smiled at the suggestion, and she thinks individuals will someday be using such data-generated personality profiles in areas like career planning. Certain personality traits are correlated with success in different occupations. But IBM's immediate plans are to apply Zhou's technology to the corporate marketplace. When we talked, IBM had three pilot projects under way. It was analyzing hundreds of millions of tweets by hundreds of thousands of people to figure ways to make marketing, customer service, and employee hiring more tailored and effective. The smart software performs a kind of alchemy, converting personal data on social networks from random writing to a digital window into a person's psyche. "Without the technology to analyze the data, it's useless," Zhou notes. "Now, it's getting to be valuable."

In September 2014, Zhou left IBM to start her own company. The idea, she says, is inspired by the work she did at IBM, and researchers there will continue to pursue the underlying technologies she developed in service of corporations. But Zhou has her eye on the consumer market. If data is the new oil, she says, then we are all data wells, and potentially valuable ones. The data-infused profiles of a person's traits and values, Zhou says, should be exploited by the individual as a kind of currency in exchange for truly personalized products, services, and advice from businesses, with tailored pricing as well. Even a prototype was months away when we spoke just after she departed from IBM, but her ambition is to help alter the terms of trade in digital commerce. "My passion is to directly affect consumers," she says, and "to really disrupt the balance of power."

D ata is power. Or it certainly can be, when animated by intelligent algorithms to generate knowledge and new services. The social and economic payoff for consumers has been enormous. The windfall includes the broad Internet utilities like Google, Facebook, and Amazon, for discovering, sharing, and acquiring, and services that enhance the bargaining power of consumers including price-comparison websites like Kayak, NexTag, and Shopping.com. Yet in this new data economy, the ultimate power resides with those who gather the data and write the algorithms.

The future will be a high-stakes balancing act. Around the world, policy makers, industry executives, and privacy advocates are wrestling with the question: What is the right balance to strike? How do you maximize the technological payoff and minimize the privacy risk?

There is no definitive answer. But by now, there are some identifiable camps of thinking—each with a somewhat different emphasis. One camp—think of it as the enlightened business community—contends that the focus of privacy rules should be on the *use* of data rather than the *collection* of data. Data, in this view, is an asset, the currency of the information economy. So, like money, the greatest value will be created if it flows freely.

That case was forcefully made by the World Economic Forum, in a report issued in 2013, *Unlocking the Value of Personal Data: From Collection to Usage*. The report grew out of a series of workshops on privacy, attended by government officials and privacy activists, as well as business executives. The corporate members, more than others, shaped the final document. Its thesis is that curbs on the use of personal data, combined with new privacy-protecting tools, can give individuals control of their own information and yet allow data markets to grow and prosper. "There's no bad data, only bad uses of data," says Craig Mundie, a former senior Microsoft executive, who is a member of President Obama's Council of Advisors on Science and Technology.

Consumer and privacy advocates are skeptical of use-only restrictions. "I don't buy the argument that all data is innocuous until it's improperly used," says David Vladeck, a law professor at Georgetown University and former director of consumer protection at the FTC. He offers an example: You spend a few hours looking online for information on deep-fat fryers. You could be looking for a gift for a friend or to research a report for cooking school. But to a data-sniffing algorithm, tracking your clickstream, your curiosity about fryers could be read as a telltale signal of an unhealthy habit—a data-based prediction that could make its way to a health insurer or potential employer.

Technologists tend to combine policy recommendations with technical fixes. Alex Pentland is leading a group at the MIT Media Lab that is working on a number of personal data and privacy programs and real-world experiments. He espouses what he calls "a new deal on data" with three basic tenets: you have the right to possess your data, to control how it is used, and to destroy or distribute it as you see fit. Personal data, according to Pentland, is indeed like modern money—digital packets that move around the planet, traveling rapidly but needing to be controlled. "You give it to a bank," he explains, "but there's only so many things the bank can do with it." The Obama administration's big-data report in 2014 renewed its call for giving consumers more control over their data, along the lines Pentland suggests. The report also said the focus of regulation should be mainly on the use of data rather than on its collection. Europe favors stronger limits on data collection.

Developing privacy-preserving tools for managing data is a flourishing niche of computer science research. The effectiveness of new rules may well hinge on the potency of new tools. Several approaches are being pursued. A promising one is "differential privacy," a method pioneered by Cynthia Dwork, which allows an analyst to ask questions of a data set and get answers without having direct access to the data. At Princeton, Arvind Narayanan is leading a project that seeks to reverse engineer what marketers do with big data to eventually create a "census" of corporate online privacy and discrimination practices. Narayanan calls it a "web transparency project," inspired by the recognition that companies often have a huge advantage in information and power over consumers in the data economy. "That's a bad thing," he says.

Data auditing techniques could also be important. In David Vladeck's example of an online search for a deep-fat fryer, an audit trail would ideally detect an unauthorized use by, say, a health insurer. Marc Rotenberg, executive director of the Electronic Privacy Information Center, prescribes a stronger dose of transparency, opening up the technology. "The algorithms should be made public," he says. "People have a right to know how the decision is made." At present, the smart code that delivers answers and offers is "a black box with consequences," Rotenberg insists.

Rotenberg's suggestion sends people in the technology industry into conniptions. The algorithms of big data are wrapped in patents and flanked by lawyers. So the chances are slim to none that the software code itself would be forced into the public domain. But take a less literal view of Rotenberg's stance, and it is not so far-fetched. More broadly, his question is: What is the recipe? That is precisely the approach being taken by IBM with a technology it calls WatsonPaths. The decision-tracking software, for example, shows physicians at the Cleveland Clinic the path of inferences and assumptions Watson made in assigning a high probability to a particular diagnosis. The software does that by showing the doctor of graphic picture of the program's step-by-step progression to a suggested diagnosis. This approach is what some call "algorithmic accountability," and what the computer scientist Danny Hillis describes as the "storytelling" that will make decisions made by artificial intelligence acceptable to society.

S o rules, tools, and social expectations will all be part of the long-term adaptation to the technology we call big data. Along the way, people will make all sorts of personal choices. I ask Ed Felten of Princeton, who knows as much about data-tracking practices and computer security as anyone, if he takes steps to protect his privacy. Occasionally, he replies. For example, he had been watching the television series *Breaking Bad*, about a high school chemistry teacher who becomes a methamphetamine dealer. Felten has done some Web searching, out of curiosity, for background on the show and meth dealing. He has read up a bit on how to make crystal meth, what it looks like, and how it is sold. For that, he uses a different Web browser and puts it in "anonymous" mode. He doesn't want to leave that digital trail. "I don't want that in my browsing history," Felten explains.

Personally, I've come to the view that credit cards are a far more accurate tracking device than my Web browsing habits. What you buy and where you buy it creates a portrait of what you *do* rather than what you *search* for. Analyzing credit card transactions along with GPS location information from cell phones is what Alex Pentland of MIT calls "reality mining," a particularly powerful form of data mining. So when I see people putting all of their daily purchases on plastic, I see people trading away their privacy for credit-card reward points. I put my annual health club membership on a credit card, but at the neighborhood wine and liquor shop I buy with cash.

Yet what Felten and I are doing are small tactical steps to frustrate some facet of the data-gathering industry. Do such things really make any difference? Probably not. You can avoid some tracking mechanisms, but trying to become a privacy survivalist seems a fool's errand. As a practical matter, there is no opt-out from the big-data world. Nor would most of us want to.

11

THE FUTURE:

DATA CAPITALISM

As towering historical figures go, Frederick Winslow Taylor was deceptively slight. He stood five feet nine and weighed about 145 pounds. But the trim mechanical engineer was an influential pioneer of data-driven decision making, an early management consultant whose concept of "scientific management" was widely embraced a century ago on factory floors and well beyond. Taylor applied statistical rigor and engineering discipline to redesign work for maximum efficiency; each task was closely observed, measured, and timed. Taylor's instruments of measurement and recording were the stopwatch, clipboard, and his own eyes. The time-and-motion studies conducted by Taylor and his acolytes were the bedrock data of Taylorism.

Viewed from the present, Taylorism is easy to dismiss as a dogmatic penchant for efficiency run amok. Such excesses would become satirical grist for Charlie Chaplin's *Modern Times*. But in its day, scientific management was seen as a modernizing movement, a way to rationalize work to liberate the worker from the dictates of authoritarian bosses and free the economy from price-fixing corporate trusts.

Taylorism was embraced by some of the leading intellectuals

of the Progressive Era. In 1910, Louis Brandeis, "the People's Lawyer" and future Supreme Court justice, wrote, "Of all the social and economic movements with which I have been connected, none seems to me to be equal to this in its importance and hopefulness." Ida Tarbell, the journalist whose investigative pieces in *McClure's* magazine helped pave the way for the breakup of the Standard Oil trust, was another champion of Taylor's techniques. Taylorism was going to replace hunch and habit with scientific precision. In his classic biography of Taylor, Robert Kanigel writes that Taylorism promised "not just science, but science wrapped in the flag of mathematics."

For a time, the reach of Taylorism seemed boundless. In *The Principles of Scientific Management*, published in 1911, Taylor wrote that his system could be "applied with equal force to all social activities; to the management of our homes; the management of our farms; the management of the business of our tradesmen, large and small; of our churches, our philanthropic institutions, our universities, and our government departments." But there was an overreaching hubris both to the man—whom Kanigel describes as often "tactless" and "pugnacious," and as someone "cruelly ill-suited to the arts of compromise"—and to his method, with its rigid insistence that measurement and analysis would yield the "one best way" to do work.

W ill data-ism prove to be a digital age version of Taylorism? There is certainly a danger that modern data analysis is misused and abused. Taylorism was a good idea taken to excess in its single-minded pursuit of one goal: labor efficiency. Modern history is filled with examples of the myopic peril of focusing on one data measurement—body counts in the Vietnam War, crime statistics in some police departments, and quarterly earnings in the corporate world. Essentially, people game the system to hit the desired numbers. This kind of behavior even has its own "law," dating back to the 1970s and named after Donald T. Campbell, a social psychologist. "The more any quantitative social indicator is used for social decision making," he wrote, "the more subject it will be to corruption pressures and the more apt it will be to distort and corrupt the social processes it is intended to monitor."

That is a pitfall, for sure. But it is created not by data but by poor management. Part of the promise of big data is that it opens the way both to more fine-grained measurement and to a broader, more integrated look at an organization's operations. Ideally big-data technology should widen the aperture of decision making rather than narrow it.

Since the Industrial Revolution, new counting and communications technologies have transformed the structure and management of business. In the late nineteenth century, railways, the telegraph, telephone, and accounting gave rise to large national corporations. By the early twentieth century, Standard Oil, General Electric, United States Steel, and DuPont were industrial behemoths. The rise of such large-scale enterprises, run by battalions of salaried managers, is what the historian Alfred D. Chandler Jr. called managerial capitalism.

Since then, the practice of managing people, corporate structure, and strategy has continuously adapted to advancing technology and changing intellectual fashions. The management of people has been a blend of elements of scientific management, which relies on incentives, control and monitoring, and humanistic management, which assumes that workers are creative and self-motivating. Managers trained in economics tend to be the scientific management camp, while industrial psychologists and organizational behavior experts tend to be in the humanistic management camp.

In structure and strategy, trends embraced over the years have included conglomerates, virtual corporations, globalization, reengineering, and strategy based on technology "platforms" rather than products. Management progress and fads have been accompanied, and sometimes triggered, by shifts in computing. The era of the mainframe, a centralized command-and-control technology, saw the proliferation of conglomerates in the 1960s and 1970s, while the personal computer and later the Internet brought reengineering, globalization, and the start-up economy.

Yet the guiding metric of management for decades has remained the same: finance. With improvements in communications and computing, financial performance can be measured faster and in greater detail than before. But it is still a fairly blunt, crude measurement—counting money, one factor of production and output of the enterprise. It is still financial capitalism. But big data promises to usher in a new phase in the practice of management. In the reporting for this book, I have repeatedly asked corporate managers and business school professors the question: Is data the "new finance," that is, a yardstick that is useful and powerful, but also can be brittle and blinkered, distorting behavior and incentives shortsightedly?

The replies I heard were much the same. A fair point, they said, but the answer is no, because big-data measurement and analysis is *qualitatively* different from the measurement of finance. They used different words and phrases to try to capture the difference. Erik Brynjolfsson of MIT's Sloan School of Management speaks of a "quantum change," while Michael Haydock of IBM talks of a new "genomics of business." In practical terms, managers talk of finance being largely a comment on the past—what has happened in the last three to six months—and mostly based on data generated inside the company for management and control. With all the new sources of information from outside the corporate walls, big data, they say, is an information-based look at the present and into the future, helped by artificial intelligence software.

It won't happen overnight, but the center of gravity in business decision-making will swing toward data. Many people I interviewed agreed with John Calkins, president of programming at AMC Theatres, and an MBA who earlier worked for McKinsey, Hollywood studios and IBM. Management, Calkins says, will become "less a finance exercise and more a data exercise." Financial capitalism will give way to data capitalism.

If data rules, the people whose skills and talents are most attuned to the new reality look like rulers. That seems to be the assumption in higher education and in the job market. America's elite universities may give elegant lip service to the humanities, but the so-called STEM disciplines (science, technology, engineering, and mathematics) are in the ascent. Students, with encouragement from their parents, are in step with the quantitative times. At Stanford, 90 percent of the undergraduates take an introductory computer science class, regardless of their major.

Here I've focused mainly on business and science, but every field is becoming data-infused these days, including the social sciences and humanities. At Harvard, Gary King made an informal count of the number of data "methods" courses being offered across all of the undergraduate and graduate departments. Methods courses, he explains, are ones that teach students how to design data experiments and do data analysis—that is, ways of learning about the world from data. King counted about 100 such courses. "It's crazy," he observes, with an approving smile. Harvard is grooming plenty of quants. Yet by 2014, Stanford had become the "it" school by some measures that Harvard long dominated— lowest undergraduate acceptance rate, most often named by high school seniors as their "dream college" in the *Princeton Review*'s annual surveys, and most money raised from donors. Harvard apparently can't compete with Stanford's location in Silicon Valley and its reputation as a hothouse of tech start-ups.

Students everywhere are increasingly preparing for a future in which jobs beckon for those with data skills. A few years ago, the McKinsey Global Institute, the research arm of the consulting firm, did a projection of the big-data job market in the United States alone. America, according to the McKinsey estimates, will need 140,000 to 190,000 more workers with "deep analytical" expertise and 1.5 million more data-literate managers, whether retrained or hired. Of course, those numbers may not prove precisely accurate. Still, the McKinsey study convincingly made the case that the data job market will be a sizable one. Yet the study, implicitly, makes another point as well. The data scientists—the men and women with "deep analytical" skills—may be the advance guard but it is the ground troops—the data-literate managers—that will really determine the pace of progress toward a data-driven society. The demand for the data-smart managers, by McKinsey's reckoning, will be about tenfold that for the quants themselves. The far larger group, in business and other fields, might be called DOPes (short for data-oriented people).

Jeff Hammerbacher and his data scientist peers have become prized commodities in the new labor market, and justifiably so. But others leading successful data efforts—Nick Dokoozlian at Gallo Winery, Timothy Buchman at Emory medical center, Donald Walker at McKesson and Menka Uttamchandani at Denihan hotels—are not young data scientists in their twenties and thirties, but rather an older generation of leaders and managers who grasp the power of modern data and analysis. They are data-oriented people, bringing big-data technology and the mind-set of data-ism into the mainstream.

The pace of big data's march into the mainstream is by no means certain. To say that something is inevitable is not to say that it is instantaneous. As I write this, the market for big-data technologies is growing fast, about 30 percent a year, and predicted to reach $24 billion by 2016, according to IDC, a research firm. But economic cycles rise and fall, and markets with them. In technology and business, revolutions tend to play out in evolutionary steps, over time. Technical innovation is only one piece of a puzzle that includes affordability, acceptance in the marketplace, and changes in behavior. Recall that nearly all of the bold predictions made in the late 1990s about the disruptive impact of the Internet across industry really did come true—a decade later, long after the Internet stock-market bubble had burst.

All successful technologies raise alarms and involve trade-offs and risks. In ancient times, fire could cook your food and keep you warm, but, out of control, could burn down your hut. Cars pollute the air and cause traffic deaths, but they have also increased personal mobility and freedom, and stimulated the development of regional and national markets for goods. The outlook for the technology we call big data is not fundamentally different. Its advance is probably inevitable, the risks seem manageable, and the benefits, by adding a layer of data-driven intelligence to the physical and digital worlds, could be as transformational as the automobile or the Internet in the long run.

Still, there are qualms. The ever-smarter algorithms of big data can be seen as the new power brokers of society, determining what information we see, products we're offered, and life opportunities we're presented. In the age of big data, we will increasingly hand over decisions to automated algorithms, knowingly or not. The real issue is under what terms we let the big-data algorithms take over—as the artificial intelligence expert Danny Hillis explained earlier, about the need for an audit trail, the need for the smart system to explain how it arrived as its software-generated decision, which he calls "telling its story."

The future of big data may well look a lot like the movie *Her*, released in December 2013 and directed by Spike Jonze. This isn't some top-of-the-head notion on my part. A number of computer scientists I spoke with mentioned it. In the movie, the big-data-of-the-future incarnation is intelligent software, fed with all kinds of information, and delivered via smartphone—one owned by the protagonist Theodore Twombly (Joaquin Phoenix). The software assistant Samantha speaks (Scarlett Johansson's voice) as his helpful assistant and eventually more. It finds the answers to questions, reads all of Theodore's e-mail and text messages, and just generally seems to know everything about him—his personal history, preferences, and tastes, the books he's read, movies he's seen, the goods he's purchased.

Samantha is a dialogue system, as they say in artificial intelligence circles, meaning it is conversational, and uses those conversations to acquire information and develop its knowledge base—machine learning on steroids. "That's where we're headed," says Larry Smarr, founding director of the California Institute for Telecommunications and Information Technology. "Hyper-personalized assistance is going to be common in ten years." The movie *Her* is set in 2025.

Is the prospect of that kind of data-driven artificial intelligence cool or creepy? For most of us, it probably seems a bit of both. It will be a while before a technology that is so human-like arrives, if it does. A digital answer machine offering remarkably personalized advice? Sure. An intelligence that is barely distinguishable from human? I wonder. A few years ago, the *New York Times* published a series of articles on the progress in artificial intelligence under the rubric, "Smarter Than You Think." The technology is impressive, and increasingly so. But what struck me while reporting these stories, and what came up repeatedly in conversations with artificial intelligence experts, is what awesome things the human brain and what we call general human intelligence really are. The general intelligence involves the effortless capacity to tap life experience, and make intuitive connections and quick decisions—what Daniel Kahneman calls "thinking fast." Then there is the human brain as a processor, cramming incredible computing power into a tiny space and using only 20 watts of energy. By contrast, the Watson computer that won its *Jeopardy!* contest with human champions burned 85,000 watts.

Still, the virtuous cycle of more and more varied data and smarter and smarter algorithms, written by human programmers, is delivering a big-data-fueled renaissance in artificial intelligence. But the more machine learning can do, the more humanity may learn about itself. "What is actually intrinsically human?" Smarr asks. "In the next couple of decades, this technology will increasingly force us to confront that issue."

NOTES

1: How Big Is Big Data?

1   Just outside Memphis: Information for the paragraphs on McKesson come from several sources. An interview on Oct. 25, 2013, with Donald Walker. An interview on Nov. 6, 2013, with Kaan Katircioglu, an IBM research scientist. An interview on Nov. 13, 2013, with James Kalina, a client executive in IBM's services group. And a technical paper, by five IBM researchers: Kaan Katircioglu, Mary Helander, Youssef Drissi, Pawan Chowdhary, Takashi Yonezawa; and two McKesson researchers, Robert Gooby and Matt Johnson. "Supply Chain Scenario Modeler: A Holistic Executive Decision Support Solution," *Interfaces* (a journal published by INFORMS, a professional society for operations research and management sciences) 44, no. 1 (February 2014): 85–104.

2   In Atlanta: Information for the Emory University Hospital comes mainly from Dr. Timothy Buchman, and interviews with him on Sept. 1 and Oct. 9, 2013.

3   David Brooks, my colleague: David Brooks column, "The Philosophy of Data," *New York Times*, Feb. 5, 2013, p. A23.

4   Ninety percent of all of the data: The estimate comes from IBM Research. http://www-03.ibm.com/systems/storage/infographic/storwize-data.html.

4   In 2014, International Data Corporation estimated: This is from yearly report on data conducted by the research firm IDC, and sponsored by the data storage company, EMC. http://www.emc.com/about/news/press/2014/20140409-01.htm.

6   The lecture was widely read: C. P. Snow's lecture was later published as a book, *The Two Cultures and the Scientific Revolution* (Cambridge

University Press, 1959). http://sciencepolicy.colorado.edu/students/
envs_5110/snow_1959.pdf.

6   Kemeny and Kurtz saw the rise: The description of BASIC at Dartmouth comes from an article by John G. Kemeny and Thomas E. Kurtz, "Dartmouth Time-Sharing," *Science* 162, no. 3850 (Oct. 11, 1968): 223-228. A book by the same two, *Back to BASIC: The History, Corruption and Future of the Language* (Addison-Wesley, 1985). And two interviews with Thomas Kurtz in 2001.

6   Years later, Gates fondly recalled: From an interview on June 22, 2001.

9   "We are betting the company": Interview on Feb. 19, 2014, with Virginia Rometty.

2: Potential. Potential. Potential.

13   Jeffrey Hammerbacher is trying to win converts: He was speaking to a group at Mount Sinai on Nov. 14, 2012. Unless otherwise noted, the quotes from Hammerbacher come from a series of interviews, both in New York and San Francisco, from November 2012 to April 2014.

15   a "paradigm shift" in medicine: An interview on Nov. 1, 2013, with Dennis Charney.

15   a "revolution" that is just getting under way: An interview on Jan. 31, 2012, with Gary King.

15   "There is a war in every field": An interview on Oct. 16, 2013, with Gary King.

15   "a transition on a par with the invention of writing or the Internet": From an article by Alex Pentland, "The Data-Driven Society," *Scientific American* 309 (October 2013): 78–83.

16   a school paper he wrote as a seven-year-old: I was given a copy of the original.

17   His father, Glenn, a factory worker for General Motors: The descriptions of Hammerbacher's upbringing and family life come from my conversations with Jeff and a lengthy interview on Oct. 14, 2013, with his parents, Glenn and Lenore. The quotes from Glenn and Lenore come from that interview.

21   One friend was Rachana Shah: The descriptions and quotes from Rachana Shah Fischer come from two interviews, on Sept. 6 and Sept. 10, 2013.

25   "books are his avenue for learning": An interview on Aug. 12, 2013, with Halle Tecco.

25   T-shaped people: This is a concept Jim Spohrer of IBM has mentioned to me repeatedly in recent years, in the context of the changing require-

ments for technology professionals. His blog post on Oct. 1, 2013, lays out his thinking. http://service-science.info/archives/3225.

26   a "typical Jeff" project: An interview on Sept. 6, 2013, with Andrew Smeall.

28   "He's very extroverted": An interview on Aug. 13, 2013, with Adam D'Angelo. Quora.

29   "a mixture of sales and winnowing": An interview on Aug. 15, 2013, with David Vivero.

3: Bet the Company

35   "a quivering mass of protoplasm—a human being": The descriptions and quotes from Dr. Timothy Buchman come from three interviews with him, on Sept. 1 and Oct. 9, 2013, and on May 19, 2014.

36   "all this technology. It's just too much": Interview on Oct. 9, 2013, with Stephanie Pieroni.

39   "Yes, it's going to be challenging in health care": Interview on Oct. 9, 2013, with Sharath Cholleti.

40   a digital form of "constant vigilance": Interview on Oct. 9, 2013, with Nagui Halim.

41   "A Business Intelligence System": Hans Peter Luhn's paper was published in the *IBM Journal of Research and Development* 2, no. 4 (October 1958): 314–19. http://altaplana.com/ibmrd0204H.pdf.

42   "most solutions can be found in the roots of math": Virginia Rometty's descriptions and quotes in this section come from an interview on Feb. 19, 2014.

46   "sweet Jesus, the inmates are going to run the asylum": William Pulleyblank's descriptions and quotes come from two interviews, on Aug. 26 and Aug. 28, 2013.

48   "What is IBM?": The description of the Smarter Planet campaign comes mainly from internal IBM documents and an interview on Oct. 1, 2013, with Jon Iwata.

52   "IBM gins up the demand, and we get the sales": Interview on March 27, 2013, with Jim Goodnight.

54   "completely disrupt the industry as we know it": Interview on Feb. 14, 2013, with John Kelly.

55   a "rocky time" for IBM: Interview on April 23, 2014, with Virginia Rometty.

56   not just a code horse, but a bit of a clothes horse: Kerrie Holley's descriptions and quotes are mainly from an interview on Aug. 14, 2013.

58   "We helped each other out": Interview on Sept. 18, 2013, with Sue
      Duncan.

59   "How many Kerrie Holleys are there": Interview on Sept. 16, 2013,
      with Arne Duncan.

                                                        4: Sight and Insight

61   "it's all about numbers, really": Interview on Nov. 6, 2013, with Kaan
      Katircioglu.

62   helped by IBM's industry consultants: Some of the description of set-
      ting up the joint project with McKesson come from an interview on
      Nov. 13, 2013, with James Kalina, an IBM client executive.

63   see its business far better at two levels: An interview on Oct. 25, 2013,
      with Don Walker.

64   inventory levels for the costly drugs were cut in half: Data on efficiency
      improvements comes from the research paper, "Supply Chain Scenario
      Modeler: A Holistic Executive Decision Support Solution," cited in
      the first endnote, first chapter.

68   more optimistic about human prediction: The Good Judgment Project
      is described on its website, http://goodjudgmentproject.com.

68   weakness so severe that she couldn't walk: Dr. Herb Chase's descrip-
      tions and quotes come from an interview on Sept. 4, 2013.

70   an estimated 10 to 15 percent of all diagnoses are mistaken: The 10
      to 15 percent estimate comes from a 2008 study by E. S. Berner and
      M. L. Gruber, "Overconfidence as a Cause of Diagnostic Error in
      Medicine," *American Journal of Medicine* 121 (2008): S2–S23. Others
      estimate that the numbers are somewhat higher, at 15 to 20 percent,
      including incomplete and erroneous diagnoses.

70   "Watson or something similar": An interview on Oct. 28, 2013, with
      Dr. Martin Kohn.

71   an intellectual champion for the transformative power of big data: I've
      interviewed Erik Brynjolfsson several times over the years, but most of
      the quotes and descriptions in this section come from an interview on
      Oct. 17, 2013.

75   detailed survey data from 179 large companies: Their research was
      published in 2011, in a paper titled, "Strength in Numbers: How Does
      Data-Driven Decisionmaking Affect Firm Performance?" The paper
      was published in the *Proceedings of the International Conference on Infor-
      mation Systems*, ICIS 2011, Shanghai, China, Dec. 4–7, 2011. Associ-
      ation for Information Systems 2011. ISBN 978-0-615-55907-0. It was
      also published online as part of the Social Science Research Network's

working paper series. http://papers.ssrn.com/sol3/papers.cfm?abstract_id=1819486.

76   a "start with the facts" approach: Brooke Barrett's descriptions and quotes come from an interview on Nov. 18, 2013.

77   A crisp, fast-talking Indian: Menka Uttamchandani's descriptions and quotes come from two interviews, on Oct. 31 and Nov. 4, 2013.

78   skepticism at first: Brian Gehlich's descriptions and quotes come from an interview on Dec. 13, 2013.

81   a start-up, Duetto Research: Some of the information on Denihan's plans for the future and Duetto's work with the hotel chain comes from an interview on Nov. 13, 2013, with Patrick Bosworth, chief executive of Duetto.

81   "We're working on it": Interview on Oct. 31, 2013, with Thomas Botts.

5: The Rise of the Data Scientist

85   "That sucks": From a *Bloomberg Businessweek* article on April 14, 2011, "This Tech Bubble Is Different."

86   including the work of Duncan J. Watts: Hammerbacher mentioned a few of Watts's writings including *Six Degrees: The Science of a Connected Age* (W. W. Norton & Company, 2003).

87   powerful signal that can identify one's spouse or romantic partner: The paper by Lars Backstrom, a Facebook researcher, and Jon Kleinberg, a Cornell computer scientist, is titled, "Romantic Partnerships and the Dispersion of Social Ties: A Network Analysis of Relationship Status on Facebook." It was published as a working paper on Oct. 24, 2013, and presented at the ACM Conference on Computer-Supported Cooperative Work and Social Computing, Feb. 15–19, 2014. http://arxiv.org/pdf/1310.6753v1.pdf.

89   "It was Jeff's intuition": Interview on Aug. 13, 2013, with Adam D'Angelo.

89   first hire was Itamar Rosenn: Itamar Rosenn's descriptions and quotes come from an interview on Aug. 15, 2013.

91   it pursued a larger goal: Jeff Rothschild's descriptions and quotes come from an interview on Nov. 20, 2013.

95   His first talk: I watched a video recording of the Berkeley lecture, stored on a university website that was password protected.

96   "The Future of Data Analysis": John Tukey's paper was published in the *Annals of Mathematical Statistics* 33, no. 1 (1962): 1–67. It is available online: http://projecteuclid.org/download/pdf_1/euclid.aoms/1177704711.

97   called an "action plan": William Cleveland's paper, titled "Data Science: An Action Plan for Expanding the Technical Areas of the Field of Statistics," was published in the *International Statistical Review* 69, no. 1 (April 2001): 21–26. It is available online: http://www.stat.purdue.edu/~wsc/papers/datascience.pdf.

98   "the great unifier": Interview on Nov. 7, 2013, with Edward Lazowska.

100   "What we lovingly call a data scientist today": Philip Zeyliger's descriptions and quotes come from an interview on Aug. 13, 2013.

6: Data Storytelling

103   "Everything you added": Interview on Sept. 17, 2013, with Sam Adams.

104   Consider start-up ZestFinance: Douglas Merrill's descriptions and quotes come from an interview on Oct. 30, 2013.

105   fees paid by payday borrowers: For the average borrowers and amounts outstanding, Merrill did his own current estimates as of late 2013. But he also referenced the Federal Deposit Insurance Corporation's report in September 2012, *2011 FDIC National Survey of Unbanked and Underbanked Households.* http://www.fdic.gov/householdsurvey/2012_unbankedreport.pdf.

    And he also referred to a July 2012 report by the Pew Charitable Trusts, *Payday Lending in America: Who Borrows, Where They Borrow, and Why.* http://www.pewtrusts.org/en/research-and-analysis/reports/2012/07/19/who-borrows-where-they-borrow-and-why.

106   people like Tara Richardson: Her descriptions and quotes come from an interview on Dec. 5, 2013.

108   Google's algorithms stumbled: The article in *Nature* by Declan Butler, "When Google Got Flu Wrong," *Nature* 494 (Feb. 14, 2013): 155–56. http://www.nature.com/news/when-google-got-flu-wrong-1.12413.

108   "big data hubris": The paper by David Lazer, Ryan Kennedy, Gary King, and Alessandro Vespignani, was published in *Science* 343 (March 14, 2014): 1203–5. http://gking.harvard.edu/files/gking/files/0314policyforumff.pdf.

108   In a follow-up paper: The updated paper to take account of the 2013–14 flu season was titled, "Google Flu Trends Still Appears Sick." It is available online: http://gking.harvard.edu/files/gking/files/ssrn-id2408560_2.pdf.

110   Never-Ending Language Learning system: I first wrote about NELL in 2010, as one article in a series of stories the *New York Times* did on advances in artificial intelligence, under the rubric "Smarter Than

You Think." That article, "Aiming to Learn as We Do, a Machine Teaches Itself," ran in the Science section, on Oct. 5, 2010, p. D1. I've stayed in touch with Tom Mitchell and NELL's progress since. http://rtw.ml.cmu.edu/rtw/.

111   "when Watson no longer answered Wonder Woman": Jennifer Chu-Carroll made that remark in a panel discussion after a preopening performance of the play, *The (Curious Case of) Watson Intelligence*, on Dec. 4, 2013. The play is about three "assistants" named Watson— the assistants to Alexander Graham Bell and Sherlock Holmes, and IBM's Watson. It was a finalist for the Pulitzer Prize in drama in 2014.

112   "The ideal of identifying causal mechanisms": *Big Data: A Revolution That Will Transform How We Live, Work and Think* by Viktor Mayer-Shönberger and Kenneth Cukier (Houghton Mifflin Harcourt, 2013), p. 18.

112   Not everyone agrees: Richard Berner's descriptions and quotes come from an interview on July 24, 2013.

113   strong clues to hidden risk bombs in the system: The paper by Emmanuel Abbe, Amir Khandani, and Andrew Lo, titled "Privacy-Preserving Methods for Sharing Financial Risk Exposures," was published in the *American Economic Review*: *Papers & Proceedings* 102, no. 3 (May 2012): 65–70. http://www.princeton.edu/~eabbe/publications/AKL_AER.pdf.

114   "Measurement Without Theory": The paper was published in the *Review of Economics and Statistics* 29, no. 3 (August 1947): 161–72. http://elaine.ihs.ac.at/~blume/koopmansres.pdf.

114   Few people have wielded the power of data: David Ferrucci's descriptions and quotes come from interviews on April 11, 2013, and Jan. 25, 2014.

115   "Why ask Why?": The paper was published in November 2013 as a National Bureau of Economic Research working paper, "Why ask Why? Forward Causal Inference and Reverse Causal Questions." http://www.nber.org/papers/w19614.

116   "The Unreasonable Effectiveness of Data": The paper by Alon Halevy, Peter Norvig, and Fernando Pereira was published in *IEEE Intelligent Systems* (March–April 2009): 8–12. https://static.googleusercontent.com/media/research.google.com/en/us/pubs/archive/35179.pdf.

116   "But to be clear": Peter Norvig's blog post. http://norvig.com/fact-check.html.

117   "Man-Computer Symbiosis": Licklider's essay was published in the *IRE Transactions on Human Factors in Electronics*, vol. HFE-1

(March 1960): 4–11. http://groups.csail.mit.edu/medg/people/psz/ Licklider.html.

117   Yet there is another view: Murray Campbell's descriptions and quotes come from an interview on Aug. 21, 2013.

119   The book, published in 2014: *The Second Machine Age* (W. W. Norton & Company, 2014) fleshes out and refines an e-book *Race Against the Machine* by the same pair in 2011.

120   "wonderful place for data scientists to experiment": An interview on Feb. 1, 2013, with Claudia Perlich.

120   "a storyteller": Danny Hillis's descriptions and quotes come from a talk he gave at IBM's Watson lab on Oct. 2, 2013.

7: Data Gets Physical

123   "This is autumn in the vineyard": Nick Dokoozlian's descriptions and quotes come from two interviews, on Oct. 15 and Nov. 19, 2013.

124   a global population of 9.6 billion people by 2050: The report from the United Nations, Department of Economic and Social Affairs, Population Division, on June 13, 2013, titled *World Population Prospects: The 2012 Revision, Key Findings and Advance Tables* (Working Paper no. ESA/P/WP.227).

128   Dokoozlian was on a mission: Hendrik Hamann's descriptions and quotes come from an interview on Dec. 19, 2013.

134   identify the next wave of technology: Jeffrey Immelt's descriptions and quotes come from an interview on Nov. 7, 2012.

134   GE tapped an outsider: William Ruh's descriptions and quotes come from interviews on Nov. 20, 2011, Oct. 30, 2012, and Oct. 2, 2013.

135   "The social impact is a big part of it": Interview with David Cronin on Oct. 30, 2012.

135   "That was a big draw": Interview on Nov. 8, 2012, with Sharoda Paul.

137   "When the machines can learn": Anil Varma's descriptions and quotes come from an interview on Oct. 30, 2012.

137   "an exciting place for data and analytics right now": From an e-mail he sent on June 16, 2014.

138   a report coauthored by its chief economist Marco Annunziata and Peter Evans: The report, titled *The Industrial Internet@Work*, was published on Oct. 28, 2013. https://www.ge.com/sites/default/files/ GE_IndustrialInternetatWork_WhitePaper_20131028.pdf.

138   "Is U.S. Economic Growth Over?": Robert Gordon's critique was a National Bureau of Economic Research working paper, published in August 2012. http://www.nber.org/papers/w18315.pdf.

139   Gordon responded to his detractors: His piece in the *Wall Street Journal*, titled "Why Innovation Won't Save Us," was published in a weekend edition, Dec. 22–23, 2013, p. C3. http://faculty-web.at.northwestern.edu/economics/gordon/WSJ_121222.pdf.

8: The Yin and Yang of Behavior and Data

143   Yoky Matsuoka was known as a robot wizard: Matsuoka's descriptions and quotes come mainly from an interview on Nov. 18, 2011.

143   Nest was cofounded by Tony Fadell: I did an article on Nest when it introduced its first thermostat in October 2011. But the definitive account of Nest's founding was by Steven Levy, published online by *Wired*, titled "Brave New Thermostat: How the iPod's Creator Is Making Home Heating Sexy," Oct. 25, 2011. I've talked to Fadell several times in recent years, but his descriptions and quotes here, unless otherwise noted, come from two interviews, on May 8, 2012, and Nov. 13, 2013.

146   "I had to live a double life": The program aired on July 16, 2008, and a transcript is available on the PBS website. http://www.pbs.org/wgbh/nova/tech/yoky-matsuoka.html.

149   Mark Malhotra, a young Stanford-educated engineer: His descriptions and quotes come from an interview on Nov. 21, 2013.

152   "Big data is the next stage": An interview on May 10, 2012, with Randy Komisar.

153   In another industry, Michael Haydock: His descriptions and quotes come from interviews on Sept. 9 and Sept. 28, 2013.

155   VALS, for "Values, Attitudes, and Lifestyles": A good description of the history and development of so-called psychographics in marketing can be found in an article by Daniel Yankelovich and David Meer, titled "Rediscovering Market Segmentation," *Harvard Business Review* 84, no. 2 (February 2006): 122–31. http://www.viewpointlearning.com/wp-content/uploads/2011/04/segmentation_0206.pdf.

159   Veronica Vargas, a young IBM consultant: Her descriptions and quotes come from an interview on Sept. 5, 2013.

161   *Models Behaving Badly*: And the subtitle is a cautionary reminder: *Why Confusing Illusion with Reality Can Lead to Disaster, on Wall Street and in Life* (Free Press, 2011).

9: The Long Game

165   Both were pitchers on the baseball team: The information on Jeff Hammerbacher's friendship with Steven Snyder and Snyder's descent

into mental illness comes partly from a talk Hammerbacher gave in 2011 at a TEDX conference in Silicon Valley. His talk, titled "Understanding Health Problems Begins with Sharing Data," is posted online at http://www.tedxsv.org/?page_id=1223.

165   "but not with the same girl": From one of my interviews with him, on Oct. 7, 2013.

168   When she met Hammerbacher: Halle Tecco's descriptions and quotes come from two interviews, on Aug. 12 and Dec. 9, 2013.

172   Schadt, like Hammerbacher, is a native Midwesterner: Eric Schadt's descriptions and quotes come from interviews on May 28 and Nov. 7, 2013.

172   Schadt is stocky yet energetic: Eric Schadt is both a character and an iconoclast in medicine. In an article in *Esquire* in April 2011, Tom Junod profiled Schadt before he came east, in a magazine piece that really captured the man and his mission. http://www.esquire.com/features/eric-schadt-0411.

174   Charney is a dapper academic: Dr. Dennis Charney's descriptions and quotes come from an interview on Nov. 1, 2013.

175   Michael Linderman, a young computer scientist from Stanford: Some description of the programming challenges at Mount Sinai comes from an interview with him on Nov. 7, 2013.

176   "this is the future of health care": Patricia Kovatch's descriptions and quotes come from an interview on Nov. 7, 2013.

179   hospital infections are a huge problem: The death estimate comes from a Centers for Disease Control and Prevention study by Monina Klevens et al., "Estimating Health Care-Associated Infections and Deaths in the U.S. Hospitals, 2002," *Public Health Reports* 122, no. 2 (March–April 2007): 160–66. The cost estimate comes from a 2013 article by Eyal Zimlichman et al., "Health Care-Associated Infections: A Meta-analysis of Costs and Financial Impact on the US Health Care System," *JAMA Internal Medicine* 173, no. 22 (Dec. 9–23, 2013): 2039–46.

180   Alex Rubinsteyn had just completed his PhD: His descriptions and quotes come from an interview on Dec. 19, 2013.

180   Tim O'Donnell was doing broadly similar research: His descriptions and quotes come from an interview on Dec. 24, 2013.

10: The Prying Eyes of Big Data

183   the Kodak-wielding "camera fiend": Information on the Kodak section comes largely from the online essays, written by David Lindsay, that accompanied the PBS American Experience series, *The Wizard*

of Photography, which aired in 2000. http://www.pbs.org/wgbh/amex/eastman/peopleevents/index.html.

184   "really freaked people out": An interview with Daniel Weitzner on March 20, 2013.

185   "not to a solution, but to an accommodation": Edward Felten's descriptions and quotes come from an interview on Sept. 24, 2013.

187   Alessandro Acquisti and Ralph Gross, computer scientists at Carnegie Mellon: Their research paper, "Predicting Social Security Numbers from Public Data," was published in the *Proceedings of the National Academy of Sciences* 106, no. 27 (July 7, 2009): 10975–80. http://www.pnas.org/content/early/2009/07/02/0904891106.full.pdf+html.

188   A Senate staff report in December 2013: It was titled, *A Review of the Data Broker Industry: Collection, Use, and Sale of Consumer Data for Marketing Purposes*. A majority staff report by the Office of Oversight and Investigations, Committee on Commerce, Science and Transportation, published on Dec. 18, 2013. http://www.commerce.senate.gov/public/?a=Files.Serve&File_id=0d2b3642-6221-4888-a631-08f2f255b577.

190   my *Times* colleague Natasha Singer: The Acxiom description comes mainly from two *Times* articles done by Natasha Singer, who has done the best reporting on data brokers and superb reporting on other data privacy issues. "Mapping, and Sharing, the Consumer Genome," published on June 17, 2012, p. BU1, and "A Data Broker Offers a Peek Behind the Curtain," on Sept. 1, 2013, p. BU1.

192   Obama administration's big data report: The policy review was led by John Podesta and the report was titled, *Big Data: Seizing Opportunities, Preserving Values*, published on May 1, 2014. http://www.whitehouse.gov/sites/default/files/docs/big_data_privacy_report_may_1_2014.pdf.

193   Latanya Sweeney, a computer scientist at Harvard: Her paper, "Discrimination in Online Ad Delivery," was published in the *Communications of the ACM* 56, no. 5 (May 2013): 44–54. It is also available at: http://dataprivacylab.org/projects/onlineads/1071-1.pdf.

197   "both fairly transparent and transparently fair": The quote and background from Michael Schrage's essay, "Big Data's Dangerous New Era of Discrimination," published on the *Harvard Business Review* blog on Jan. 29, 2014. http://blogs.hbr.org/2014/01/big-datas-dangerous-new-era-of-discrimination/.

197   Zhou is a diminutive dynamo: Michelle Zhou's descriptions and quotes come from interviews on Nov. 18, 2013, and on Jan. 14, 2014.

200   in the spring of 2014, Zhou presented a paper: The paper was titled, "KnowMe and ShareMe: Understanding Automatically Discovered

Personality Traits from Social Media and User Sharing Preferences," by Liang Gou, Michelle Zhou and Huahai Yang, *Proceedings of the SIGCHI Conference on Human Factors in Computing Systems*, pp. 955–64. The conference was in Toronto, April 26–May 1, 2014.

203   That case was forcefully made: The World Economic Forum report, titled *Unlocking the Value of Personal Data: From Collection to Usage*, was published in February 2013. http://www3.weforum.org/docs/WEF_ IT_UnlockingValuePersonalData_CollectionUsage_Report_2013.pdf.

203   "There's no bad data": An interview on Feb. 26, 2013, with Craig Mundie.

203   "I don't buy the argument": An interview on March 15, 2013, with David Vladeck.

204   "You give it to a bank": An interview on Oct. 17, 2013, with Alex Pentland.

204   a method pioneered by Cynthia Dwork: A technical overview of the approach is described in her April 2008 paper, "Differential Privacy: A Survey of Results." http://research.microsoft.com/pubs/74339/ dwork_tamc.pdf.

204   seeks to reverse engineer what marketers do with big data: Arvind Narayanan's descriptions and quotes come from an interview on March 3, 2014.

205   "The algorithms should be made public": An interview on March 2, 2014, with Marc Rotenberg.

11: The Future

207   Frederick Winslow Taylor was deceptively slight: Information and quotes from Taylor's writing come from Robert Kanigel's *The One Best Way: Frederick Winslow Taylor and the Enigma of Efficiency* (Viking, 1997). It is the definitive biography of Taylor and his times.

209   what the historian Alfred D. Chandler Jr. called managerial capitalism: Chandler won the Pulitzer Prize in history for his chronicle of the rise of modern business management up to the 1970s, *The Visible Hand: The Managerial Revolution in American Business* (Belknap Press/ Harvard University Press, 1977).

211   "less a finance exercise and more a data exercise": An interview on Sept. 10, 2013, with John Calkins.

211   At Stanford, 90 percent of the undergraduates: From an interview on Nov. 22, 2013, with Bernd Girod, professor of electrical engineering and senior associate dean for online learning and professional development at Stanford.

211   "It's crazy": Gary King's descriptions and quotes come from an interview on Oct. 16, 2013.

211   Stanford had become the "it" school: Based on article in the *New York Times*, "To Young Minds of Today, Harvard Is the Stanford of the East," by Richard Pérez-Peña, published on May 29, 2014, p. A1.

212   a projection of the big-data job market: The McKinsey Global Institute study was titled, "Big Data: The Next Frontier for Innovation, Competition, and Productivity," by James Manyika et al., May 2011. http://www.mckinsey.com/insights/business_technology/big_data_the_next_frontier_for_innovation.

214   "That's where we're headed": Larry Smarr's descriptions and quotes come from an interview on May 22, 2014.

INDEX

cameras, early privacy concerns and, 183–86
Cameron, William Bruce, 10
Campbell, Donald T., 209
Campbell, Murray, 117–19
cancer therapies, genomics research and, 179
Carnegie Mellon University, 110–11
Chandler, Alfred D., Jr., 209
Charney, Dr. Dennis, 15, 174–75
Chase, Dr. Herbert, 68–70
Cheever, Charlie, 86
chess. *See* Deep Blue supercomputer, of IBM
Chester, Jeffrey, 189–90
Choletti, Sharath, 39
Chu-Carroll, Jennifer, 111
Cle-Metric, 39
Cleveland, William S., 97
Climate Corporation, 129
cloud computing, IBM and, 9, 54
Cloudera, 83–84, 99–101
computers
    early privacy concerns and, 184–85
    human experience and intuition and, 117–21
conceptual computing, 136–37
context
    and collaboration between computers and humans, 117–21
    correlation and, 109–17
Cope, Oliver, 37
correlation, of data, 103–4
    context and, 109–17
    context and relationships between computers and humans, 117–21
    Google Flu Trends and, 107–9
    ZestFinance and, 104–7
Cronin, David, 135
Cukier, Kenneth, 112
Cutting, Doug, 92–93

Dalio, Ray, 114
Dandy, Walter, 37
D'Angelo, Adam, 28, 88–89
Dartmouth College, 6–7
data
    data auditing concept, 204–5
    data-first ethos, 33, 96
    measurement and, 5
    quantitative-to-qualitative transformation and, 7–8
    sources and quantities of, 3–5
    see also correlation, of data
data brokers, privacy concerns and, 188–92
data-ism
    data capitalism possibilities, 213–15
    origin of term, 3
    pitfalls of, 207–9
data science
    in academia today, 15–16, 97–98, 211–12
    origins of, 95–97
David, Paul, 72
Davis, Dr. Kenneth, 174
decision making
    big data and, 5, 10, 209
    Denihan Hospitality Group and, 76–81, 212
    "fast" and "slow" thinking and, 66–68, 215
    human experience and intuition's, relationship to data, 64–76
    McKesson pharmaceutical distribution center and, 61–64
Deep Blue supercomputer, of IBM, 40, 117–19
Deming, W. Edwards, 10
Denihan Hospitality Group, 76–81, 212
Derman, Emanuel, 161
D. E. Shaw Research, 100, 180
differential privacy concept, 204

Icahn Institute for Genomic and Multiscale Biology. *See* Mount Sinai Hospital

Imbens, Guido, 115–16

Immelt, Jeffrey, 134, 141, 142

industrial Internet, of GE, 133–37
potential economic benefits of, 137–42

"Information, Technology and Information Worker Productivity" (Brynjolfsson, Aral, and Van Alstyne), 74

information gatekeepers, identifying of, 74

Intelligence Advanced Research Projects Activities (IARPA), 68

International Data Corporation, 4

"Is U.S. Economic Growth Over?" (Gordon), 138–40

Iwata, Jon, 48–52

Jelinek, Frederick, 110

Jennings, Ken, 40, 114

Jeopardy, 7, 40, 110, 114

Jobs, Steve, 64–66

Kahneman, Daniel, 66–67, 71, 112, 215

Kanigel, Robert, 208

Kasparov, Garry, 40, 117–18

Katircioglu, Kaan, 61–62

Keene, Kevin, 156

Kelly, John, 54

Kemeny, John, 6–7

Kepler, Johannes, 116

Kim, Heekyung, 75

King, Gary, 15, 211

Kleinberg, Jon, 88

Kleiner Perkins Caufield & Byers, 144, 151–52

KnowMe software, privacy concerns and, 197–202

Kodak cameras, early privacy concerns and, 183–85, 186

Kohn, Dr. Martin, 70–71

Komisar, Randy, 151–52

Koopmans, Tjalling, 114, 116

Kovatch, Patricia, 176

Krugman, Paul, 139–40

Kuhn, Thomas, 175

Kurtz, Thomas, 6–7

Kwan, Nesita, 189

Landsat satellite, 130

Lazowska, Edward, 98

Lenovo, 54

Licklider, J. C. R., 117, 119

Linderman, Michael, 175–76

Lindsay, David, 184

Lo, Andrew, 113

Luhn, Hans Peter, 41, 95

Lundgren, Terry, 161

machine-learning software, 5, 30
GE and, 133–42

Macy's, data science and, 157, 158–62

Malhotra, Mark, 149–50

management trends, of the past, 209–11

"Man-Computer Symbiosis" (Licklider), 119

marketing, uses of big data, 195–97

Maslow, Abraham, 155

Matsuoka, Yoky
background of, 143, 145–47
Nest learning thermostat and, 143, 145, 147–49

Mayer-Schönberger, Viktor, 112

McAfee, Andrew, 119–20, 139

McGregor, Dr. Carolyn, 40

McKesson pharmaceutical distribution center, 1–2, 3, 9, 61–64, 79, 212

McKinsey Global Institute, 212

ABOUT THE AUTHOR

S teve Lohr reports on technology, business, and economics for the *New York Times*. He was a foreign correspondent for a decade and served brief stints as an editor before covering technology, starting in the early 1990s. In 2013, he was a member of the team that was awarded the Pulitzer Prize for explanatory reporting. Steve has written for magazines such as the *New York Times Magazine*, the *Atlantic*, and the *Washington Monthly*. He is the author of a history of computer programming, *Go To: The Story of the Math Majors, Bridge Players, Engineers, Chess Wizards, Maverick Scientists and Iconoclasts—the Programmers Who Created the Software Revolution*. He lives in New York City.